PRAISE FOR DAVID SHEFF'S *GAME OVER*

"Beguiling . . . almost as hypnotic as a successful video game."

—*New York Times*

"A cross between *Barbarians at the Gate* and *The Soul of a New Machine*."

—*Chicago Tribune*

"Finally, a book as provocative as its title. *Game Over* is a detailed, fascinating, and instructive case study."

—*Fortune*

"*Game Over* . . . is ultimately less absorbing than 'Tetris' but not by much."

—*Wall Street Journal*

"For business moguls who someday want to corner their markets, this book is a must-read. . . . *Game Over* is about as readable as a business book can be."

—*Houston Chronicle*

"Writing with the playful pluck of Mario, the little protagonist of Super Mario Bros., Sheff unfolds an engrossing tale."

—*People*

"An absorbing read, even for non-Nintento junkies."

—*Birmingham News*

"My advice is simple: Read Sheff's book."

—*Wichita Eagle*

CHINA DAWN

★

The Story of a Technology and Business Revolution

DAVID SHEFF

HarperBusiness
An Imprint of HarperCollins*Publishers*

For Tiger Feng, my godson
And for Karen Barbour, a candle in the darkness

Some material in this book first appeared in a
different form in *Wired* magazine.

HarperCollins books may be purchased for educational, business,
or sales promotional use. For information please write: Special
Markets Department, HarperCollins Publishers, Inc., 10 East 53rd
Street, New York, NY 10022.

FIRST EDITION

Designed by William Ruoto

Library of Congress Cataloging-in-Publication Data has been
applied for.

ISBN 0-06-000599-8

02 03 04 05 06 ❖/QW 10 9 8 7 6 5 4 3 2 1

Enlighten the people generally, and tyranny and oppression of body and mind will vanish like evil spirits at the dawn of day.
—THOMAS JEFFERSON

With our technology, enlightenment can flow through the taps like water.
—EDWARD TIAN

Qihu nanxia.
(Once you get on the back of a tiger it's difficult to dismount.)
—CHINESE EXPRESSION

CONTENTS

At dawn, the coast road is shrouded in fog. On the horizon, the sun pierces the brume like a signal lamp on a great ship. My friend Bo Feng and I drive for an hour and park at a roadside dirt lot where we load up our backpacks and head out. Surfboards under our arms, we trudge to a secret spot on the California coast. It's a spectacular beach with blowing white sand dunes, playful harbor seals, and dive-bombing pelicans. The ocean comes alive here when the wind is offshore and the swell is strong and from the south. Were I to reveal the location, my surfer friends would think nothing of drowning me. They would be justified in doing so.

We pad along a brushy deer trail that leads to the beach and then a mile more in the sand. At land's end, we change into four-millimeter-thick wet suits, charge into the ocean, and paddle on our boards in the icy surf until we are past the breakers, where we await a set of waves. They arrive with respectable size and power. Held aloft by a steady off-shore wind, they break in long, peeling curls.

China has no equivalent of wave surfing—no beach breaks, no beach culture. So of course it was only after going to the West that Bo began to surf. Some would say that it was predestined. In Chinese, *Bo* means "wave"; in classical Chinese, *Feng* means "to walk on the water in bare feet."

When I met him a decade ago, Bo was a busboy, waiter, dish-washer, and sushi chef working at Chinese and Japanese restaurants in Marin County across the Golden Gate Bridge from San Francisco. In the swiftest and most unforeseeable career change since Jesse Ventura's, he is now an investment banker and venture capitalist, funding entre-preneurs in his native China. The entrepreneurs are the founders of a diverse group of information-technology start-ups with a common pur-pose: They are devoted to building the Internet on the Chinese Mainland, where the majority of people are without telephones, never mind access to the Net. When he talks about his work, it's with sweep-ing emotion and momentous themes. "My country is simultaneously going through what America went through in the 1900s and in the 1990s—industrializing, building an infrastructure, beginning to trans-form from a rural economy, and at the same time racing into the world economy based on communication and technology," Bo explains. "Both highways—literally, the roadways, as well as the information superhigh-way—are being constructed at the same time."

Bo is a tall and glamorous figure who intrigues almost everyone who meets him. He's mercurial—now warm and open, now contempla-tive and impenetrable. You don't want to miss what he has to say. His intelligence is curving, circumnavigating ideas and attacking them at unexpected angles. He observes the big themes in everyday events. You want to go where he goes.

Sitting in the green water with a sea otter that is nibbling its breakfast (the meticulously plucked-off legs of a still moving crab), I watch Bo as he readies for the lead wave in a set. Paddling—slowly at first, then hard and even—he leaps atop his board and cuts up the wave's face. After a quick turn, he slides down, his board shooting out a spray of white water.

The parallel to Bo's life strikes me. Bo surfs on the water between China and the United States, and it is his ability to navigate the vast ocean that separates our two nations that makes him a leader of the latest revolution in China—a revolution with the potential to transform the life of more than a fifth of the world's people. Few ride seamlessly between our world and his, but Bo makes it look easy. If only he surfed as well . . . As I watch, cringing, Bo slips, and his board shoots off in one direction while his body topples down the wave's face. The breaking water pounds him. Just as he swims to the surface and gasps for air, a larger wave thrashes him. The metaphor has become more exact. Surfing as he does between China and America is fraught with peril. Conditions change without warning. One could easily be caught inside and swallowed up. Yes, one could drown.

Before I accompany him to China for the first time, I think I know Bo pretty well. In China, however, I realize that I only know half of him. The other half isn't exactly different, it's just more: more speed, more urgency. The poles of his mood are exaggerated, reflecting urban China at this particular juncture in history. It's one of the most vibrant places on the planet, where each day has a life-or-death sense of purpose, despair, frustration, opportunity, dread, and hope. The only other place and time I have felt anything comparable was when I visited Moscow in 1987. I had been invited to participate in Soviet general secretary Mikhail Gorbachev's historic Peace Forum. There were Nobel Prize winners, clergymen, economists, physicists, global business leaders, artists, and writers. Gorbachev himself addressed us. So did Andrei Sakharov. Gore Vidal, sitting near me, observed, "Democracy has come to the Soviet Union. The surest proof: At our meeting at the Kremlin, Claudia Cardinale entered the room and not one man in the place stood up to give her a chair." Norman Mailer said, "The degree of openness in the Soviet Union is incredible. The last thing one expects to find is that the USSR is a place vibrant with hope and passion." We saw it outside the conference halls on the streets of Moscow and other cities we visited, where people—Soviet citizens in train stations, on buses, on crowded market streets—were exuberant with expectation. On the famed Arbat in central Moscow, a group of students agreed to pose for a photograph in front of a statue. Before I clicked the shutter, an elderly woman rushed forward, her arms waving. "No make photo!" she yelled. "No make photo! It is prohibited!" Then one of the students movingly stood up to her, speaking solemnly, her eyes aflame. "No," she said. "No more! It is allowed. We are free. It is Gorbachev time." Gorbachev had recently announced a new era of glasnost and perestroika, intended to liberalize and revitalize Soviet society, and the nation's optimism (however short-lived it turned out to be) was palpable.

So, too, is the hopefulness in China. It's the last thing I expect to find. Bo notwithstanding, I arrive in China with a collection of bleak preconceptions. A tyrannical government and unsmiling people in drab clothing: stoic and inscrutable. Executions doled out like parking tickets. Children laboring in sweatshops. Unprovoked arrests and the black hole of the Chinese gulag. The routine infanticide of female children. The ominous and violent spring of 1989, when tanks rolled into Tiananmen Square and soldiers fired upon China's own citizens, murdering hundreds, maybe thousands of people. However, after a tumultuous and grim century and a decade after the devastating and demoralizing Tiananmen Square massacre, the people I meet in China seem to hold nearly uncontainable optimism about the latest revolution that promises to transform their nation. Neither violent nor velvet, it is a virtual revolution—that is, a revolution at the subatomic level of electrons. Digital packets and beams of light are invisibly but profoundly transforming China. How? For generations and more than a century, information and communication in China were restricted and censored. Opportunities for individuals to determine their own lives—where they lived, whom they married, their jobs—were nearly unheard of. Now, as Bo says, "The carp is leaping through the Dragon's Gate." The Chinese expression describes the transformation of a poor man into a rich one because of his hard work, but Bo recasts the idea. The efforts of the Chinese entrepreneurs are propelling China forward. The Internet is transforming the nation. As a result, China herself will leap through the Dragon's Gate and become something new. And back to Russia: The Chinese people I meet exude confidence that this revolution is unlike the Soviet one that led to the disastrous collapse of the Russian economy. The Chinese revolution, they maintain, in spite of setbacks and formidable opposition, is cautious, scalable, and socially and economically responsible. Bo expresses this repeatedly. "*Rou hu tian yi*," he says, quoting another gnomic saying. It means, "It's like adding wings to a tiger." The idea is a transfiguring force that is so powerful as to be almost mythic. Bo says, "With the Internet we are adding wings to a tiger."

Bo and his friends are an unlikely group of revolutionaries, yet their similar histories brought them together and lead to their crusade. They are businesspeople born at the time of the Cultural Revolution and raised on Mao, scattered by Deng Xiaoping's reforms and then sobered and politicized by Tiananmen Square. The trauma of June 4, 1989, inspired them to repatriate and found businesses with a mission. (That is, businesses with a mission beyond the making of money, though that's allowed.) Bo says, "In America, you are taught that you can control your own destiny. That was not the message in China when we were children. In China, we never thought we could make a difference. Now I think maybe I can." Bo and his friends are committed to information technology, particularly the Internet, precisely because of its potential to change their world.

I have heard similar pronouncements before. Every CEO of every entrepreneurial company I've ever met in California's Silicon Valley has described his or her company as revolutionary: *"Our company will change life as we know it!" "We are changing the world!"* Whereas such innovators as Andy Grove, Steve Jobs, Tim Berners-Lee, and the like are true pioneers of the information revolution, most company founders I heard from were doing little more than providing alternative ways to buy books, flirt, or unload Fiestaware. The grandiosity of their ambition seems to be a consequence of the need for justification, in a culture that reveres idealism and demands a higher purpose, for people who have neither. But the Chinese business leaders I came to know—a small group of China's most important information pioneers—are different. In most cases, they could have more comfortable lives in America. However, they are devoted to building the Internet and other communications technology specifically to shake China loose from its stagnant, isolated, and repressive past, to raise the standard of living, and create new-economy jobs. (Bo says, "If we don't develop a vibrant IT economy, China will do little more than continue to make shoes for Nike.") They believe that technology can vastly improve China's educational system, a necessary predecessor of change. Most important, the entrepreneurs are passionate about building an indestructible, modern infrastructure that includes an uncensored and uncensorable pipeline of free-flowing information that will connect China—first its teeming metropolises and eventually its remote villages—with the rest of the world.

Is it realistic or naive? "Sweeping change is not only possible, it's a certainty," argues a Shanghainese entrepreneur. "However, if our goal is importing a just, representative form of government, we have to remember the process by which a democracy comes into being. Imposing democracy is oxymoronic. You can't force democracy from the outside. The information revolution is setting the stage for radical change. The seeds are planted. What will grow? Democracy? A more responsive CCP [Chinese Communist Party]? Something else entirely? We don't know." We don't know what happens when technology transforms the business, social, and political life for a fifth of the world's population and China comes racing at light speed to catch up to the West. Of course there are enormous opportunities for Western businesses to cash in as China becomes the world's second largest (eventually, possibly the largest) technology customer, but the China market is the least significant part of the revolution. "What happens when China successfully transforms from a mainly agrarian/industrial nation into one that has significant input from the information technology industry?" asks Jay Chang, an analyst who covers China for Credit Suisse First Boston. "What happens when eighty percent of the state-owned enterprises in China are able to link economically to the global Internet on fast pipes? What happens when China's engineering talent pool is able to gain access to high-end computing resources and exchange ideas and information easily with their global peers? What happens when fifty percent of the Chinese population gets wired in ten years—six hundred million people, the largest number of Internet users in the world?" Only time will provide the answer, but one thing is certain: It's impossible to overstate the seismic nature of the coming changes. "China is not just another player," says Lee Kuan Yew, the senior minister of Singapore. "It is the biggest player in the history of man."

In China, I feel the explosive combination of forces aligning to create the kind of change that alters the course of history. They draw me into this story about the revolution and some of the most prominent revolutionaries. It is a fundamentally different undertaking for me. Throughout my twenty-five years reporting and writing, I have maintained a distance from the subjects of my articles, interviews, and profiles in order to maintain my objectivity as a journalist. This book is an exception. It would never have been conceived without my close friendships with Bo and his wife, Heidi Van Horn. Introduced by Bo, I also become a close friend of Edward Tian, one of the revolution's most dynamic leaders, and his family, who for a year live near my family's home in California. During this time, our children attend the same school and play together. Remarkable conversations with Bo, Edward, and their colleagues inspire me to write a series of magazine articles, and then this book, about them and their work. Over the course of the three years I spend researching and writing, I become more involved. In 2000, I agree to become an unofficial adviser for and investor in Chengwei, the venture capital fund Bo founds with a partner. My motivation: our friendship and a belief in the work that he and his friends are doing.

These relationships mean that I have access and gain insights that would be impossible for other journalists. My Chinese friends insist that few Westerners, whether businesspeople or journalists, come to understand the people of China even as foreigners are arriving by the planeloads. Guidebooks instruct businesspeople to go out drinking with their Chinese counterparts in order to get to know them, but it takes patience, perseverance, and flexibility, not cocktails or gifts, to get things done in China. Moreover, it takes time, openness, and sensitivity to foster trust.

The friendships that develop mean that I have the extraordinary opportunity to come to know, with a remarkable degree of intimacy, some of the men who are working to transform China. There is no guarantee that they will succeed; in fact, the odds are against them. As things play out, however, it is as if one journalist had the opportunity to witness and chronicle the emergence of key leaders of the U.S. technological revolution before they had such a profound impact on our culture. I am along for the ride, recording how these visionary entrepreneurs and venture capitalists fare through the evolving political climate; governmental regulations that change weekly; the roller-coaster investment climate; boardroom coup attempts; and world events such as the bombing of the Chinese Embassy in Belgrade, China's struggle to enter the World Trade Organization, the collision of a U.S. spy plane and a Chinese jet fighter, and the cataclysmic terrorist attack on the United

States on September 11, 2001. The reader should remember that the relationships that allow such access mean that I am not objective about my subjects or their work. I am a friend and participant, however peripheral, in the events I'm recounting. My bias is set from the beginning, but that doesn't mean that I report a whitewashed version of my subjects' stories. From the start they understand that my book will only be credible if I write about their difficulties—including their failures—as well as their successes. As a result, all of my subjects speak candidly and almost all conversations are on the record, though a few people ask me to protect their identities because of their fears for their or their families' safety. History has proved that in China, caution may be prudent. Rousing a sleeping dragon can be dangerous.

A note about the names in this book: In China, family names precede given names. The names of the people in this book follow the form used by the individuals when they are living or working in the West. For example, my friend's Chinese name is Feng Bo, though in the United States he goes by Bo Feng. ("Feng" is pronounced "Fung.") Tian Suning uses his adopted American name, Edward Tian. Wang Zhidong prefers the Chinese form on both sides of the Pacific.

—David Sheff, San Francisco

LIKE ADDING WINGS TO A TIGER

On a Sunday morning in early 2000, while loudspeakers on a passing advertising van pump out tinny jingles and a froggy computerized voice croaks, "Try Peace Cigarettes, fresh and fashionable," Edward Tian looks out on the arcing concrete-and-metal Lu Gouqiao Bridge on the outskirts of Beijing. "The bridge is the site of the first bullet of the Chinese-Japanese war," he says. "Sixty years ago, the Japanese marched over this bridge to occupy Beijing. Now, below the bridge"—he points northward to a grassy field—"is our fiber. We never were successful fighting in the old world. We suffered greatly. But in the new world we are as strong as any country—stronger than most. This fiber"—he shakes his head, looks almost misty eyed—"is

China's reemergence, and I think that things will get better for our people."

Edward has neatly cut black hair that falls over his forehead and serene, deeply brown eyes. Built compactly, he sits erect in the back seat of a Shanghai-made Buick with his arms folded on his lap as we cross the bridge. We drive on for an hour or so through Hebei Province before the car turns up Zuan Nan, a thin *hutong,* or alleyway, in Gao Pei Dian. It's a dust-blown, sun-faded terra-cotta-and-pearl-colored village with crumbling brick houses jammed together. We park and head down a narrow dirt path between worn granite pillars with monkeys carved in the face of the stone. Chickens scatter as we walk.

Wearing a gray suit and polished dress shoes, Edward, the thirty-seven-year-old CEO of China Netcom Corporation (CNC), seems out of place in Gao Pei Dian. So does the gaggle of eager, youthful, short-sleeved and open-collared engineers who escort him, ducking under a low beamed doorway into a small shop where a woman is weighing out mounds of powdery medicines (antler, mushroom, and bark) onto cut squares of brown parchment. She gives our tromping entourage a furtive, unimpressed glance and goes back to her work.

Edward and I follow the others out the back door to another pathway that winds past bare-chested men playing a card game, a woman sweeping a stoop, and a half dozen skinny children in short pants and sandals. At the end of this path, Edward and the engineers enter the emerald-colored doorway of a fifties-era two-story concrete block building and clomp up a flight of concrete stairs. At the top there's a hall with a bed, a small wooden desk, and an ordinary wooden door.

On the other side is the kind of nondescript room you might expect to see in an abandoned government office building (exactly what this is). It smells musty from mildew and mold and the army-green paint is peeling. The light, exposed fluorescent tubes and a few dangling sixty-watt bulbs, is flickering and dim. Yet in this dank room, seeming out of place, are racks of Cisco, Huawei, and Lucent components, including DWDM repeaters with flashing diodes and blinking digital readouts, all connected by snaky yellow cables. As incongruous as the idea is, this is ground zero for the revolution. "Here," Edward says, gesturing around the room, "is where the past gives way to the future and

China transforms." He looks directly at me. "It all comes down to a word. It all comes down to bandwidth."

Edward Tian is known throughout China as the man who built, from the ground up, the nation's Internet. As a cofounder of AsiaInfo, the country's first homegrown Internet infrastructure company, he helped construct the backbone of the country's national and many provincial networks. AsiaInfo is one of the most dramatic success stories in the new China—the first private Chinese firm to go public in the West. Which is why he baffled many of his friends and angered AsiaInfo's board when he announced that he was leaving for a government-funded start-up designed to compete with the 500,000-employee state-owned monopoly, China Telecom.

Edward and the engineers are here to inspect the newest node in what will become—at twenty thousand kilometers and forty gigabytes per second—the longest and fastest high-bandwidth network in the world. As he says, "Cities that never had phone service are being wired for broadband. We are basically wiring the nation with fiber that will bring limitless opportunity to the people."

In its first year, with 750 employees and an army of 20,000 subcontracted workers, CNC has dug eight thousand kilometers of trenches—that's a couple of thousand kilometers longer than the Great Wall of China—and filled them with the cables that house fiber optics that connect China's seventeen largest cities. On the route there are 550 Optical Fiber Distribution Centers like this one in Gao Pei Dian placed every sixty to eighty kilometers along the network, since light waves traveling over the fiber must be regenerated, or amplified, after traveling that distance. (Currently located in old government buildings, these centers will soon be moved into new, ultraclean, temperature-controlled structures with redundant power supplies.) The company is already offering a wide range of products, including access to the network, Web hosting, enterprise software, datacenters, phone cards, and a range of other Internet and Intranet services. In less than a year, CNC has rolled out the depth and breadth of services that it took WorldCom and Sprint years to build. Indeed, analysts are saying that CNC has the potential to become China's WorldCom or Sprint (with IP-based voice and data communication), Level 3, Quest, Exodus, Global Crossings (and others) all rolled into one. Credit Suisse's Chang predicts that

"CNC could become the largest and one of the most influential companies in the world." In the *Wall Street Journal*, Thomas Ng, head of Venture TDF, a Singapore-based VC fund, is similarly effusive: "They may well become one of the largest companies in the world." Not just in China; both say "in the world."

Edward says that he envisions his network giving birth to an economic boom in China unlike any the world has seen—tens of thousands of start-ups as well as services that will modernize traditional companies that together will revitalize the nation's sagging economy. As a longer-term goal, and even as he acknowledges the huge obstacles, he also foresees a social revolution from radically improved systems of education and healthcare, all made possible by unprecedented access to information. Those in turn lead to—what? Though he doesn't pretend to know what it will look like, he says, "A new society. A new life for the people of China based on opportunity. A strong nation."

Edward examines a piece of transparent fiber about the width of a fishing line and asks a bespectacled engineer, "Is this G655 fiber?"

Assured that it is, Edward turns back toward me. "One pair of this fiber can carry forty gigabytes of data per second." How much is that? He responds, "Enough to carry all of the conversations going on on all of the AT&T lines at any single time in the United States." If that isn't enough, he adds, "We are already upgrading to a new type of fiber that holds sixteen hundred gigabytes per pair." It's twice the speed of WorldCom's UUNet network—the state of the art in the United States—and comparable to the fastest fiber in the States and Europe.

I do some calculating. There are ninety-six pair of fibers in place and the potential for several hundred more. That equals . . . He fills in the thought for me: "limitless bandwidth." Edward says, "Imagine what limitless bandwidth can do for China. With our technology, enlightenment can flow through the taps like water. When it does, it will enlighten our whole country."

His work at the Gao Pei Dian station complete, we follow the engineers back through the herbalist's kitchen and jump into three cars, which caravan away from the rundown neighborhood to a six-lane highway that follows a train track out of town. There is a cornfield on one side and a row of shops, garages, and storehouses on the other.

The cars stop on a roadway and our group crosses it, dodging trucks filled with sacks of rice (and a dozen men riding upon them), metal beams, and polished new Chinese-made VWs. When we make it across, we clamber up a dirt embankment that leads to a peanut field that borders the train tracks to a raised concrete dome shaded by a *liu* tree. On top of the dome is a large manhole cover, which two of the engineers pry open. (Edward is concerned. "These must be secured. We need locks on them. That should be a high priority." An engineer writes it down as an action point.) The manhole opens to a dome-shaped, concrete underground room. Eight black conduits stream in from the north-facing wall and exit through the south. Only two are connected by smaller cables—cables filled with the smaller cables that are in turn filled with the hair-thin fiber-optic strands. Edward explains that the technology reflects the world's state of the art, but it is also upgradable when new types of fiber are developed or more is needed. To add or replace fiber, a machine blows the filament one and a half or so kilometers through the tubes to the next manhole.

On the way back to the car, a train shoots by. A hundred cars. The noise, the smell, the sight: moving metal. The iron rooster was a symbol of industrial-age China. The symbol for the information age is Edward's fiber, with its invisible digital flow coursing at the speed of light. A hidden dragon. It's why one of Edward's engineers sums the mood of the people working in the Chinese IT companies when he says, "Everyday we feel as if we are writing China's history."

As his car pulls away from Gao Pei Dian, we drive by a gathering of lean, brown-skinned men stripped to the waist sitting on boxes, their backs resting against the walls of a windowless brick building, smoking cigarettes on the side of the road. Edward looks at them with deep interest. I've come to know the look on his face as one that precedes magniloquence, and he doesn't disappoint. In fact, it's the most succinct explanation of the vastness of his vision yet. Most observers still consider the potential commercial Internet market in China to be the 250 to 300 million people in large cities who earn three or more times as much as rural Chinese, but Edward says that his network will one day help the people on the side of the road, too. The other billion. "Everyone knows that there are many people in China who don't have phones, never mind computers," he says. "Yet we will bring them high-bandwidth fiber. What good is it? They may not use e-mail or e-commerce anytime soon, but ubiquitous bandwidth will affect their lives when we wire not only the cities but the villages—the clinics, schools, and libraries. A new China will emerge. Since the broadband Internet can be picture and voice run, illiterate people can have access, perhaps coming to them on the eight hundred million televisions in their homes. One day every school will be on the network. For people who were previously presented with one alternative for their lives, broadband brings a vision of other choices and the means to reach them. When we can bring state-of-the-world education to every child in China, our children will grow up as good as anyone." In an unwavering voice, Edward explains that Chinese children often die because of inadequate healthcare in poor regions like Hebei. "Given time, broadband may even be able to help change that," he says. He envisions the Internet as a tool that will one day help train local doctors and connect village clinics to medical centers in big cities. "Today," he says, "a child is born with something as simple as a cleft palate and nothing is done—he becomes a ghost." For more advanced consultations, specialists in the medical centers could see patients and read their vital signs in real time.

Where the fiber reaches major cities, the network is being connected to high-bandwidth rings that link up major buildings. Two hundred fifty buildings are targeted in the first wave. Through the buildings, CNC is offering a wide range of high-bandwidth, IP-based net services with voice, 3-D imaging, teleconferencing, fax, as well as data communication. Further in the future, Edward will connect the network directly to similar networks in North America and Japan with a broadband transcontinental submarine cable. It's a $300 million proposition, but Edward says it can be accomplished in only nine more months. "The undersea cable will be a blood vessel connecting two people," he says. "It will be less likely that misunderstandings will be allowed to fester if we are connected by a single blood vessel. It means a more stable world, a stronger one. There is much less anger when you can see a person's eyes."

In China, one discovers that there is a saying for almost every life circumstance. "Paper cannot wrap up a fire" means that the human spirit is irrepressible; the truth will reveal itself. It can seem quixotic in China, where the government has a history of concealing the truth and employing intimidation and brute force to quell the human spirit. China is changing, however.

There is a newsletter and magazine that is widely read in China that is distributed solely on the Internet. Called *Dacankao, (VIP Reference)*, the periodical is sent, unsolicited, to more than a million e-mail addresses throughout the Mainland. It's written and edited in Washington, D.C., by Li Hongkuan, formerly a professor at the Medical Center at Beijing University, who goes by the name Richard Long in the United States.

Since Long and his fellow editors and writers are overseas, they are beyond the reach of the Chinese security forces, and it would be unreasonable to punish the readers of *Dacankao*, since they receive the newsletter without requesting it. The head of Shanghai's computer security police has reportedly received a copy in his "IN" basket. The website of the vibrant periodical, at www.bignews.org, is plastered with crimson banner ads and satiric photos of Chinese leaders. A reader on the mailing list on the Mainland says that it includes "All the news that Beijing deems unfit to print," including editorials critical of the Communist regime and essays by dissidents and links to their websites. Neither is *Dacankao* the only Internet-based newspaper that reports stories, including investigative articles and wide-ranging opinions, that would otherwise go unpublished in China, where the Communist Party exercises rigid control of the press.

One investigative report that appeared in the proliferating online journals covered the fatal stabbing of a seventeen-year-old girl in a Shanghai hotel and the shoddy police investigation that followed. Local newspapers and television stations ignored it, but when the article was published on the Internet, the incident was discussed in cafés, on buses, on the streets—everywhere. The police, pressured by government officials who feared a scandal, reopened the investigation of the murder, a series of events unheard of in China. In August 1999, *Dacankao* ran the story and digitized pictures of an outburst of anti-Chinese violence in Indonesia—a news story banned in China because party leaders reportedly feared that it would incite demonstrations. Nonetheless, the Web-based report was passed among computer users, printed out, and circulated throughout China. As predicted, students gathered in protest outside the Indonesian embassy and may have influenced Beijing's subsequent tougher line with the Indonesian government. Between April and June 2001, China executed at least 1,781 prisoners, according to Amnesty International. A report called "Where Did My Brother's Body Go?" published in a small newspaper in Jiangxi Province described how the organs of a man executed for a multiple murder were sold without his or his family's permission. According to the *Washington Post*, "Though the Beijing government has attempted to suppress discussion of organ-harvesting, the article was picked up by the *People's Daily* Online, the main Web site of the most powerful official newspaper in China, where it remains posted." The man's sister sued, and the government reluctantly opened an investigation. There are numerous other examples of the ways that the relatively free online underground press and other Internet offerings have become thorns in the side of the Beijing government. E-mail has been the primary communication tool of the Chinese Democracy Party since it was banned in the late 1990s. The persecuted Falun Gong religious sect used the Internet to mobilize against the government's crackdown. In fact, reports of the first wave of police violence against members of the cult spread on the Net and resulted in an online call to arms among supporters. When, on April 25, 1999, ten thousand Falun Gong members gathered in front of Zhongnanhai, the governmental leadership compound in Beijing, the protest had been plotted on the Internet. Since then, Falun Gong has been banned and many of its members imprisoned in jails and mental institutions, but the group has continued to organize on the Net.

The Internet is too unwieldy and decentralized to effectively censor, but that hasn't stopped the security forces from trying. There has been an ongoing campaign to monitor content, including surveillance and filtering. Since connections with the outside world are squeezed through three government-monitored gateways in Beijing, Guangzhou, and Shanghai, the government tried programming firewalls to block pornography and "incendiary" material. However, the blocks were largely ineffective and slowed down the national network, so most were abandoned. Other attempts to restrict content are countered by hackers, who use proxy servers as gateways to sites that are officially blocked. A 2000 study by the Chinese Academy of Social Sciences in Beijing revealed that a quarter of Internet users in China use proxy servers. Richard Long says that the government has as many as thirty thousand

Internet police doing filtration, but *Dacankao* usually gets through. "How can they check thirty million e-mails a day?" In addition, Chinese Internet users increasingly employ software such as Triangle Boy, which tricks electronic filters. The program, created by a U.S. company called Safeweb, is free and widely distributed. A twenty-year-old engineer who works for a state-owned high-tech company in Beijing spoke to me on the condition of anonymity, summing the opinion of many in the hacker community in China. "You cannot control the uncontrollable," he says. "When they try to block websites, we use proxy servers. When they read our e-mail, we bounce messages off systems that they'd never check. If they close down a discussion forum, we'll create another. If they want to listen in our chats, we'll use more layers of disguise."

Some factions of the government are equally tenacious, of course. The most egregious examples of the reactive forces are a handful of arrests and prosecutions.

In November 1999, Lin Hai, a Shanghai-based software designer, sends Richard Long a lengthy list of e-mail addresses of Chinese computer users. The e-mail is intercepted and security forces show up at Lin's home. Though his wife protests that Lin is apolitical and that he was exchanging e-mail addresses with Long in order to build his company's mailing list, her husband is dragged to jail and charged with using the Internet "for the overthrow of the State."

Lin Hai goes to court in December. It's a closed-door trial, and it surprises no one when Lin is found guilty. His sentence: two years in prison. His computers and modems, "the tools of his crime," are confiscated.

Lin's arrest and sentence leads to a worldwide "cyber protest" that includes e-mail letters and an online petition to the Chinese government. As usual in China, the protest seems to fall on deaf ears, though there is speculation that the government may be increasingly susceptible to foreign public opinion because of the country's ongoing campaign to be admitted into the World Trade Organization (WTO) and its bid to host the 2008 summer Olympic Games. The Internet community doesn't let the Lin Hai case disappear. Neither does the arrest dim the commitment of the Internet's pro-democracy activists, who use Lin's story as an online rallying cry.

Some representatives of some government ministries and agencies in China try to control the Net with new regulations and arrests.

However, if it perceives the Net as a threat, why doesn't Beijing simply pull the plug? The dominant forces in the government clearly understand that the Internet and other information technologies are essential to the nation's rapidly expanding economy, integral to the country's post-Deng obsession: to return China to the position it held before the so-called century of humiliation when it was a leader in the world in art and science and its economy soared. When I ask him about this, one Chinese entrepreneur says, "China has been held back because of a century of stagnation while the world moved ahead. Here is our opportunity to catch up." In addition, with the overall Chinese economy in decline, including some sectors in crisis, the Net and related technologies are expanding. Besides the obvious effects such as providing new economy employment—essential if China is going to effectively modernize—and bringing in badly needed capital, state-of-the-art technology is essential for a wide range of Chinese businesses to be competitive in the world market. "The Net means that China can stop playing the poor relation," Bo says. "The Net levels the playing field." If nothing else, the leaders in Beijing know that China can only be a successful participant in the global economy, particularly when it joins the WTO, if the nation is wired. It's why nearly everyone involved agrees that there's no going back now that the Internet is a national policy.

By building a state-of-the-art Internet for business, Beijing has set an irreversible course, unstoppable even if the Net begins to undermine the Communist Party's rigid control. Orville Schell, renowned China scholar and dean of the Graduate School of Journalism at the University of California at Berkeley, repeats a Chinese saying. "Qihu nanxia," he says. It means, "Once you get on the back of a tiger it's difficult to dismount."

CHAPTER I

WITH RICE AND RIFLES

It is the summer of 1998, when the futuristic Capital Airport is under construction across the tarmac, so our 747 parks at old Beijing Terminal, a blocky warehouse with the unmistakable design flair of Stalinist China: emaciated green. After maneuvering through immigration, I am met by a chauffeur. As instructed, I follow him outside to a scuffed charcoal Buick. He holds open the rear door and speaks a rehearsed greeting in English: "Welcome to China. Fasten your seat belt."

Rarely have I received advice that is more prescient. The driver, like the mass of Beijing drivers, attacks the gas pedal and horn with equal glee, untroubled by lanes, one-way signs, or roadway shoulders. A small jade charm, like a glass LifeSaver, dangles on a red string from the rearview mirror. For safety. Over the course of the harrowing drive along tree-lined highways, we narrowly miss a man on a rickshaw, a van that transports people and split pork carcasses, a truck spilling green melons, and hundreds of bicyclists, including one who is balancing sheets of corrugated metal on his head.

The car speeds by a building-long sign of the times, a mural depicting a farmer with a hoe alongside a technician soldering an integrated circuit. The title painted in slanted red lettering: AGRICULTURE, TECHNOLOGY—HAND IN HAND WORKING FOR THE COUNTRY. Mottoes like this one, ubiquitous on billboards and banners, are as stilted as the slogans from the Cultural Revolution that implored the Chinese to embrace Mao Zedong Thought and denounce the imperialists (us!), "capitalist roaders," and their running dogs.

Pedestrians leaping for their lives, the car tears into the financial district of Beijing with its wide boulevards lined by buildings shimmering in the sunlight. (They are sparkly on this ultrabright, almost luminescent day. I will see them soon enough in their more normal state, clouded by grimy brown Beijing haze.) The car skids to a stop on— that's *on*—the curb of a crowded street lit up with megawatts of street lamps and blinking ads for noodles and dot-coms (CLICK AND GET SICK reads an inexplicable billboard). I jump out and duck into a traditional restaurant with red balloon lanterns with yellow tassels and a round doorway. Inside, there's Bo.

Facing three men around a cuneiform dining table in an undecorated private room, Bo sticks out among the gathering. He has a fresh buzz haircut, titanium-framed glasses, and an improbable Nike ensemble, while the others wear short-sleeved shirts, dark pants, and black shoes.

Bo introduces me. Sitting at the corner of the wedge is Edward Tian. Next to him is Liu Yadong, a slightly taller and thinner man with high cheekbones and a pinched face, a pale complexion, and a floppy pompadour. Bo says that Liu is Edward's old friend and his company's chief operating officer. The other man is Gong Hongjia, youthful in his mid-twenties with a broad face and a thick broom of jet-black hair. Gong, I'm told, is the founder of a successful start-up. He founded Dekang to market software he developed to bill telephone calls just as the wireless boom began in China. He has 30 percent of the market and sales of $6 million. "Gong Hongjia's company is the leader in its field, but there is growing competition from abroad," says Bo.

Gong looks up and half smiles. "Yes," he says. "We are fighting harder. We are fighting with rice and rifles."

Mao once said that the people of China would persevere against the far better armed forces of Chiang Kai-shek's American-backed Kuomintang army "with rice and rifles." At Gong's appropriation of the Maoism, the group laughs heartily.

"That's us fighting the West," says Gong, now solemn. "But make no mistake. Intel, HP, Microsoft, Oracle, and the others assume that China will be theirs, but they underestimate the Chinese entrepreneur."

Bo adds, "Most foreign companies hire a group of highly paid executives who live in $10,000 a month apartments in Shanghai. A lot of U.S. companies send over a sales-and-marketing team and think they'll nail China. They don't have the awareness or the drive that the Chinese entrepreneurs have. There's no comparison between entrepreneurs who run businesses in China and U.S. companies that have outposts here."

He orders dinner and we watch a young waiter stick his hand into a glass case that is teeming with vipers and "eye" snakes. He snatches one tightly below the head and carries it, its body whipping and writhing, away into a back room. The snake appears later on a platter. There are crispy bony chunks grilled and painted with sweet vinegar and shallot sauce. The snake's skin, thinly sliced, has been tossed with cucumbers in a salad. Each person is presented with a pair of tulip shot glasses filled in turn with crimson red and cloudy yellow liquids. One holds rice wine mixed with the snake's venom and the other is rice wine with its blood. Bo lifts his glass and the rest of us follow him. "To China," he toasts, and we drink down the bloody elixir, which travels simultaneously down my spine and straight to the backs of my eyes, which tear up. Bo has an enormous smile. "Now you are Chinese," he says. The others laugh uproariously and Bo says, "You'll have wild dreams tonight."

Alongside upholstered Nokias sit plates of the snake along with lobster sashimi (the lobster is still moving) and baked tortoise. Chairman Mao once said, "The Revolution is not a dinner party," but this one appears to be. It is a feast fit for heads of state, or in Bo's case one of China's first dedicated high-tech VCs.

It strikes me that this meeting would have been inconceivable only a short time ago. Venture capitalists may be a familiar breed in the West, but Bo is still a rarity in China. For most intents and purposes, Bo is like his brethren in Silicon Valley. Like them, he provides key resources—capital, advice, and leadership—to promising companies for equity and fees that, if all goes well, lead to an "exit" (read that: a killing, preferably an acquisition or an IPO on the Great Wall Street) for his investors. However, unlike most Stateside VCs, Bo speaks both as a player and a patriot. That is, he embraces venture capitalism as a means to a loftier end. "Here venture capital is the water in a fertile valley," he says. If the technology itself is adding wings to the tiger that is China, venture capital is adding wings to the tiger that is the entrepreneurial movement. With passion that borders on zealotry, he continues, "Venture capital is an engine that drives forward a social revolution." Sandy Robertson, the famed investment banker who is Bo's mentor and a longtime China watcher, says that the impact of the revolution can't be overstated. "Bo and his crowd are having much more of an impact on the new China than the old-guard leaders who preoccupy the West. They are creating the new China."

When Bo became a VC, he quickly emerged as one of the hottest in the exploding Chinese IT market. After a short year or two, Chinese government regulatory agencies, ministers in Beijing, and officials in Shanghai solicited him to educate them about Western-style venture capital. His good reputation gives him access to many of the best deals. Representatives of Morgan Stanley Dean Witter, Goldman Sachs, and other foreign banking and investment firms regularly approach Chinese start-ups with more money, says Liu Yan, a former associate of Bo's. "But the entrepreneurs are not impressed. Bo impresses them. He can speak to them."

The reason he can speak to them is telling: the specific combination of shared and distinct history that drives him and places him at the apogee of the entrepreneurial movement that is transforming China. It's also what makes him and the entrepreneurs with whom he works fundamentally different from their American counterparts. It's not just that their market is different and there is a different time paradigm in

China—insanely quicker. Neither is it just the historical context to their work. Like the founders of the greatest Silicon Valley companies, they are driven, brilliant, devoted, creative, and inexhaustible. (In fact, they may even work harder than the Americans did. Len Baker, the respected venture capitalist who funded such companies as Palm and BroadVision, says, "I have never seen a work ethic at any time in Silicon Valley like I see in China.") The biggest difference, however, is the precariousness of their lives—a degree of risk that never existed in the Silicon Valley. The reason: China herself. China is an ever-present and uncontrollable factor in every venture—meddlesome, potentially dangerous, and completely unpredictable. Regulations and rules change dramatically and constantly and the government retains a direct or indirect stake in many ventures. Says Duncan Clark, cofounder of BDA, an Internet consulting company in Beijing, "This environment—the regulations, the competition, the political uncertainties—makes these the fastest, most courageous, nimblest-thinking people globally. To deal with this level of risk and still sleep is no small accomplishment. But they're hooked on it like some Chinese are becoming hooked on Starbucks cappuccino." No, Bo and his friends couldn't be living in more extraordinary times, but the times demand a lot. They need single-mindedness and talent, razor-sharp wits and patience, and the ability to think while running. They need an iron constitution and an understanding (or no) spouse. So the significant difference between them and the Stateside VCs and entrepreneurs is their ability to tolerate—to tolerate and to thrive with the added danger. Bo and the others are navigating in uncharted waters, where two contradictory systems meet to create a third system that is neither Western nor traditionally Chinese. What rules there are in China are made up along the way and business, politics, and cultures collide on a daily basis.

After dinner, Bo, Edward, Liu, and Gong regroup at the sixteenth-floor bar of Beijing's fashionable Swissôtel. Bo informs the group that he is quitting smoking and is no longer buying cigarettes. Nevertheless, over the course of the meeting, he bums cigarettes from Liu, the waiter, the bartender, and a man at the next table.

The cigarettes are sucked in over green tea and beer. Bo writes notes on a tiny pad, scribbling answers to a long list of questions accumulated over the past six months of his investigation of Gong's business. When the conversation wraps at two in the morning, a deal is close, but not closed. Edward as well as Gong and Liu say goodnight, but it is morning in the States, so Bo begins speed-dialing New York and Sunnyvale. He retires at four in the morning.

Bo wasn't exaggerating about the dreams. In one, the waitress from the restaurant is a snake-haired Medusa with a fax printing out of the center of her skull. He wasn't exaggerating the lack of sleep, either. The alarm is bone chilling and I hit the breakfast room as if I'd been dropped through the roof. We eat a bowl of soupy rice congee and jump into a taxi, which proceeds along the Second Ring Road. We shoot past a Tibetan lama temple, the home of Maitreye, a Buddha carved from a single gargantuan tree. It took three years to transport the Buddha to Beijing from Nepal. Maitreye is the "Buddha of the Future."

We weave in and around more taxis, packed green buses, bicycles, trucks, rickshaws, and pedestrians. Then we turn down a thin, pebbled *hutong*, where gray-walled homes, some five hundred years old, line both sides. The shortcut leads to Zhongguancun, Beijing's Silicon Valley. It is as different from California's as imaginable. Rather than sprawling corporate campuses, it feels more like East Broadway in Manhattan, though here the vendors are selling circuit boards and software instead of tube socks. Zhongguancun, one main center of the city's multibillion-dollar technology industry that represents a third of Beijing's economy and houses at least a thousand technology companies, is an urban scramble of intersecting streets. Crammed amid retail shops and Chinese restaurants plus McDonald's and Dunkin' Donuts are neon-lit outlets for electronic gear that includes everything from disk drives to Sparc Stations. Beginning at the Third Ring Road, Zhongguancun heads north along Haidian Lu, the main street, where hardware companies like Apple, Compaq, and Legend, the number one PC maker in China, have offices and showrooms. Innumerable software companies are tucked along side streets, and further to the south lie myriad congested multistoried R&D centers and assembly plants. It all ends in the north at Beijing University, which spawned many of the technology companies. In Zhongguancun, you can't miss the euphoria—and the vertigo: China, the Internet, and the Net in China all represent opportunity, but risk is in the air, too.

Bo and I charge up the stairs inside an office building. We are meeting Edward Tian at AsiaInfo, where the corporate mission statement is painted in large blue Chinese characters along the entry-way hall.

TO SEIZE THE OPPORTUNITY BROUGHT BY THE INFORMATION REVOLUTION, TO REALIZE THE TECHNICAL DREAM AND TO SERVE OUR COUNTRY. THROUGH OUR GENERATION'S HARD WORK, TO SET UP AN INFORMATION TECHNOLOGY BUSINESS, USE CHINA AS ITS CORE, CREATE WORLD STANDARDS—THE BEST SOFTWARE AND SERVICE FOR CHINA TO ENTER THE ELECTRONIC AGE.

Sitting at the polished rosewood Ming-style desk in his office, Edward is peering at a computer terminal, shooting off e-mails. When he sees Bo, Edward stands up and gives him an affectionate, brisk handshake and a pat on the back. The two men sit head-to-head in facing chairs and feverishly buzz about Gong's company. Bo tells Edward that he plans to fly to Hangzhou, Gong's home base, to negotiate the particulars of the acquisition. Edward says, "We've talked enough. We've done our due diligence. It's time to make a deal."

Outside AsiaInfo's office, Bo and I are again in the backseat of the car. Bo fusses with his cell phone; the battery is dead. He borrows the driver's phone to call his wife, Heidi Van Horn, in California—one of his twice-daily calls. He speaks baby talk to his month-old son Tiger. Bo makes half a dozen more calls. If cell phones turn out to be lethal, Bo's a goner. (Who isn't? This revolution would fizzle without cell phones.)

Bo scribbles notes in miniscule script on a tiny notepad. Yes, he has offices in Beijing and San Francisco (he has plans to relocate to Shanghai), but he prefers the office in his pockets—notebooks, organizers, and cell phones. His conference rooms are in any number of restaurants, cafés, teahouses, mah-jongg parlors, hotel lobbies, bars, airport lounges, train cars, dumpling stands, gambling dens, and the backseats of taxis and limousines.

The car double-parks and we walk past small shops, a KFC, and street vendors (of buck-fifty "Nike" shoes and pants by "Pravda" and "Abercrombie and Fish" [sic]). We pop into the bamboo entryway of a teahouse and trot up rickety stairs for a quick meeting with a group of Solomon Brothers bankers stationed in Beijing and then Bo places more calls, making appointments for later in the week, including a large gathering of AsiaInfo directors for Sunday night. (The workweek is so busy in China that the only "free" time for meetings is often Sunday nights.) Back in the car, he returns the cell phone to the driver and asks him to take us to the airport.

It's evening and we fly Air China to Shanghai, where the stunning, brand-new Pudong Airport is like a set for a futuristic technothriller. The gates run for acres in an inverted wedge that is made of glass and metal spines. The main terminal is crowned by a series of grand arcs that look like enormous orchid petals. In it, security is streamlined, and we are on the road, in a shiny silver Audi, within minutes of landing.

The car roars through the Pudong District, where cylindrical office towers have sprouted tall on wheat and rice fields. Seventy-five percent of the world's tallest cranes have been working here throughout the late 1990s building apartment buildings, office towers, and hotels. This section of the Pudong is shiny and immaculate.

After crossing the river, spiraling around a cloverleaf, and ripping past the stately Bund ("Revolutionary Boulevard" during the Cultural Revolution), the car turns into Nanshi, the ancient world of old Shanghai. The car hogs the alley (called a *longtang* in Shanghai), its horn adding to the cacophony of bicycle bells and swearing and street vendors hawking watermelon and sesame-onion flatcakes. As we tear past sand-colored buildings with red roofs and peeling blue doorways, I look up and see the towering metal of the Pudong skyscrapers.

Leaving Nanshi, we arrive at the Ritz-Carlton, where Bo holds court. There are a series of meetings in the lobby amid horse and lion sculptures. Bo has sessions with entrepreneurs with whom he is con-sidering investments, ones focusing on thermal printing, CD-ROMS, global satellite positioning systems for trucking companies, enterprise and database software, and a central website for Chinese real estate.

One young man who arrives at our table seems particularly out of place in the lobby bar, where a Chinese harpist and flutist play on the mezzanine. He's Gao Limin, a thirty-one-year-old software designer with long, barely combed hair, who's dressed in a shirt by Polo Izod (sporting a polo player *and* an alligator) and jeans. His round jade eyes simultaneously greet and size up. His arms move like shorting electric wires. Over beers, he tells me that he is the son of engineers and that he worked as a software designer in China and Germany for Siemens, where he first "saw the vision" of the Internet. "Suddenly," he says, "I can know everything. Here was a taste of freedom. Every voice is on the Internet. If you can hear every voice you can learn to think for yourself. In China, this is the first step toward transformation."

Gao says that he had known that his friends in China were beginning to surf the Web in 1996. He returned to China last year and with "my comrades"—four friends—started an Internet company modeled after U.S. financial sites such as Silicon Investor and Motley Fool.

I ask the first question that almost every Westerner asks about the Net in China: Doesn't the government censor it? He says, "Yeah, they put some blocks on some sites, but they're ineffective. We know how to find whatever exists on the Net." Later, when I betray some cynicism and ask how a financial website contributes to a social revolution, he responds, "The desire to make money is what will grow the Internet in China first. That is why we choose this commercial venture. The key to this stage is building the Net. It will exist if businesses go forward. Once it exists, it will always exist. Our goal is to make the Internet strong and undefeatable."

I speak fast, but Gao speaks faster, whether about margin trading or servers, talking while devouring the free peanuts and spicy chips. Before he leaves, Bo hints that he may be interested in investing, but there's work to be done: a tight business plan and an untangling of the complex ownership structure of Gao's company. (Like many Chinese IT start-ups, this one is backed by the government, which presents a morass of complications.)

Though it's the last of the successive meetings, Bo is unable to wind down. In fact, he sort of rocks forward in the cushy Ritz chair and points the lip of his beer bottle after the departing Gao. "See that!" says Bo. "Entrepreneurs like him are the most vibrant force in China. That's why I love my job. People on the outside complain that the changes in China are coming too slowly. In the U.S., people decry the problems of human rights or the restrictions on the Net, but they don't understand China at all. What they aren't acknowledging is the miracle we're witnessing. They're complaining but we're thinking, *My God! Look at the progress!* A few years ago it was tough to get a dial tone in China. There was no information. None. Now we're the fastest-emerging player into the global market. Technology is accessible to many people. We're on the cutting edge, competitive with anywhere you can mention. Many of our schools have been disasters, but slowly the Internet will go into schoolrooms and villages. Someday they will be as good as anywhere. It is piping in opportunity. Before this, people could hope for nothing more than a two-hundred-dollars-a-month job. One day people will be able to become trained online and get a job that pays a living wage. We're going forward step by step."

In the morning, Bo and I board an Air China flight to Hangzhou, a city for which he has great affection. After the Cultural Revolution, when Bo was six years old, his family took the only vacation of Bo's childhood in Hangzhou.

Hangzhou is a robust city, picturesque, lushly colored, with universities, industry, and the famous Oriental Pearl, the West Lake. In the late-morning light, watery irises float atop its now tea-green water under a hot sun, round and orange like a halved melon. The crushing heat is bearable in the garden under the shade of willows.

We walk through the garden to our hotel, the Shangri-La, which once was a favorite of Chairman Mao. (In the lobby there's a gallery of photographs of his visits here.) The grounds are lush, covered with green-velvet lawns and manicured hibiscus and orchids. Kudzu lines a fishpond, and nearby stone-slab benches look out on the lake upon which sampans float noiselessly.

There's no work until tonight, so Bo and I jump into a banged-up cab that takes us up onto the nearby mountain, past acres of tea plants like vineyards in Napa. As instructed by Bo, the taxi stops at Longjing, or the Dragon Well, where two mineral water streams meet. "The elevation is higher," Bo says. "The water is pure. As a result, the tea is very good." Now the sky is rose-colored, streaked with deep blues, and the driver accompanies us along the mossy path to the famed well, where he grabs a broken tree branch. With it, he stirs the black well water and, smiling broadly and toothlessly, says, "The dragon is here." I look in. It *is* here, materialized in the swirling pool.

The driver sits outside and has a cigarette and Bo and I enter an old wooden hut with glazed roof tiles. The proprietress shows us to a long table and Bo asks to taste the season's best tea. When she reappears from a back room, the woman ceremoniously unwraps a brick-sized package of dried and lightly roasted tea leaves, which Bo examines—smells, runs his fingers through—and sends back. The four "virtues" of a tea are its color, smell, taste, and shape, he explains. "My father loves tea," Bo says. "He has taught me to appreciate the subtle virtues." He says that he didn't have to taste the tea he just sent back; its smell and color were not pleasing.

Better samples are brought out until Bo examines one that he wants to taste. It's a ritual in this tea shop that hasn't altered much in twelve hundred years. The woman takes a pinch of leaves and sprinkles them into lidded cups filled halfway with hot water. The leaves "wash" and float downward before more water is added. Bo, using the lid of his cup to push aside the floating leaves, tastes the pale liquid. He smiles, signals for the taxi driver to come inside, and pours glasses of tea for us all. The driver takes a sip and nods his approval. When we're finished, Bo buys the Longjing tea, "the sweetest dew of heaven."

At night, we dine at Bo's favorite Hangzhou restaurant. Another banquet. "I can work hard but the food is the reward at the end of the day," he says. Afterward, we stop and sit on a bench on the shore and look out at the now-olivaceous lake, its overgrown edges, muddy beaches, and dark bluffs with hidden houses. We board a flatboat and a burly man in a sleeveless T-shirt paddles us across the lake. Ashore, we jump into another cab and head to a smoky teahouse, where Gong Hongjia arrives with an assistant. Gong, with his blocky face and hair stuck up in a cowlick, looks sharply through the café. When Bo calls to him, Gong rushes over. "I have prepared what you asked for," he says.

Gong presents a report with answers to the questions Edward had posed in Beijing and Bo carefully pores through it. He seems satisfied with what he reads. Over more tea, the two discuss details of an AsiaInfo acquisition. Price and Gong's position in the company are the sticking points. Bo begins methodically with a volley of questions about Gong's goals and an analysis of the numbers that he has presented. The negotiations, interrupted by a steady stream of phone calls on their cells, continue until the café closes. It looks as if the deal is finally ready to close. Gong had been pushing for a prestigious position and high salary at AsiaInfo, but he now seems willing to retire from the company in order to pursue other interests, including investing in Internet start-ups. A complete buyout of Dekang won't cost extra cash, just more AsiaInfo stock.

Bo promises Gong a call in the morning and we say goodnight, leaving for our hotel. It's still hot outside at four in the morning. Bo calls Heidi in San Francisco, where it is time for lunch. "An aching comes from the distance," he tells me. "I need to hear her voice."

In the morning, Bo and I return to Beijing and taxi to AsiaInfo, where Edward is preparing for the Sunday night board meeting. Bo fills Edward in on the meeting with Gong and says, "We're there if you want it." Edward plans to confirm the deal with the board.

By Monday, it's done. There is a casual signing ceremony at AsiaInfo. Both sides seem pleased. Bo and Edward rib Gong about becoming a man of leisure. If he chooses, he can retire, particularly after AsiaInfo goes public (a year and a half later) and Gong's net worth hits about $25 million.

Gong leaves Edward and Bo, who take a moment to reflect. "It's like the Long March, but after hundreds of years of wars and revolutions, it seems like a miracle that we Chinese have a chance to build our country," says Bo. "Yes, China is a rusty machine. We have to fix it, do whatever is required."

"If we do," Edward adds, "in the future our children will have the same opportunity to dream that other children have."

CHAPTER 2

THE MYSTERIOUS ISLAND

So is man's heart. The desire to perform a work which will endure, which will survive him, is the origin of his superiority over all other living creatures here below. It is this which has established his dominion, and this it is which justifies it, over all the world.
—JULES VERNES, *The Mysterious Island*

In the West, we assume that dreaming is intrinsic to human beings, that it is part of the definition of our humanity. Everyone dreams. But Edward disagrees. He says that his life has taught him that we must learn to dream.

Edward's given name is Suning, an uncommon Chinese name that means "Remember Leningrad." His parents chose his name so that whenever they uttered it—indeed, whenever they thought of their son—they would recall the happiest time of their lives, when they met and fell in love in Russia. It was prior to the icing of relations between China and Russia in the 1960s, when the two countries were close allies. Privileged Chinese children were sent to be educated in the Soviet Union, just as they are now often sent to the United States and Europe.

Edward's parents, both biology students, met in Leningrad in 1954 at the Forestry Technology Academy. They returned to China in 1960 and were both assigned to work as researchers at the Chinese Academy of Sciences. They married in Beijing in 1961. Suning was born two years later and was sent to live with his mother's parents in Shenyang because his parents were assigned to a research institute in Lanzhou, in the geographic center of China in desolate Gansu Province. Traditional temples on the hills surround the industrial town and the port of Lanzhou, where the Lan-Xin Railway runs along the old Silk Road, linking China's most remote outposts and the populated south. The train roars from Lanzhou through the "Gate of Demons" to the Gobi ("place without water") Desert.

Liu Shu and Tian Yuzhao were assigned to different desert regions. Yuzhao's job was to research the adaptation of plants in the changing desert environments in a remote area in Xinjiang. Shu worked at Sha Bo Tou, an experimental base station, on the construction of a railway across twelve kilometers of sand dunes between Baotou and Lanzhou. Building tracks across the unstable, drifting sand of the desert was an extremely tricky problem. Her breakthrough in the field earned her a national prize.

Shu and Yuzhao rarely saw one another, but once each year they met to travel to visit Edward. When they arrived, he was cautious.

"Do you know who we are?" Shu asked at one of the reunions.

"You are Mom. You are Dad."

"He knew us," Shu says, "but he didn't have any affection for us. In his face was fear that he would have to leave my mother. When we departed, he held on to my mother's leg so tight and did not want to go to the train station to say good-bye to us. As a mother, I felt sorry about that, but because of my work and ideals, I had no other choice."

Life in Shenyang with his grandmother, the principal of a middle school in Liaoning Province, was carefree. They lived in a large house with cherry and pear trees not far from the Hunhe River. Edward was given a great deal of freedom to play in the neighborhood, and he remembers chasing dragonflies in the park and climbing trees. "My son had a very happy and vivacious childhood until he was four years old," says Tian Yuzhao.

However illogical, beginning in the 1950s Mao Zedong set his sights on teachers, along with writers, artists, scholars, and most other professionals who had been revered and privileged under the initial Communist regime, labeling them "Reactionary Bourgeois Authorities." In the late 1950s and early 1960s, Mao's Hundred Flowers and Antirightist campaigns that preceded the Cultural Revolution were the first of the purges of teachers and intellectuals, particularly singling out those who had been educated abroad. They continued during the Cultural Revolution, which began in 1965. Now that the USSR was an enemy of China, the Tians' degrees from Leningrad meant that Yuzhao and Shu were denounced as "cow's demon and snake spirit," mythical demons that assumed human forms in order to do evil. They were separated and sent to concentration camps for *laodong gaizao,* or "rehabilitation through labor."

Throughout China, Mao's citizen militia, the Red Guard, empowered by their vague mission to rid China of the "Four Olds" (old culture, old customs, old habits, and old ways of thinking), physically and psychologically tortured innocent people and destroyed incalculable amounts of private and public property. The Guard in Shenyang took over Edward's grandparents' home and dozens of people were moved in. The family was confined to a small back room. When Edward was five years old, Red Guard troops forced Edward's grandfather to surrender his beloved collection of world literature along with the family's few heirlooms to a bonfire held in the yard. Uncomprehending and panic-stricken, Edward ran up to a guard and tried to rip a book out of his hands. "These are my grandfather's books," he yelled, but the guard knocked Edward to the ground.

Once a month, grain, cooking oil, and miniscule amounts of meat and eggs were rationed. "In addition," Tian Yuzhao says, "there was not only poverty in the material aspect, but in a moral aspect, too. The people of the whole nation struggled against each other during the Cultural Revolution."

Uncertainty and terror marked Edward's early life. He heard nothing from his parents, from whom he was separated for most of a decade, for as much as a year on end. Edward, who maintains that his life and the lives of those of his generation are divided into three stages that coincide with the three stages of China's recent history, now says that the Cultural Revolution was the Dark Age of his life and the Dark Age of modern China. It was characterized for him by a single, all-pervading emotion: hunger. "All I remember is hunger," Edward says. "Hunger for food. Hunger for information."

After years of torture and humiliation, the government's investigators concluded that the Tians had clear political records and they were released from the camps and sent back to Lanzhou, though they weren't yet allowed to return to their fieldwork. Edward remained in Shenyang until his grandfather died in 1970. His grandmother couldn't care for him herself, so he was sent to live with his parents for the first time in his life. He lived with them for two years before becoming seriously ill, possibly because of the remote region's contaminated water. Edward, who was eleven years old, was hospitalized and his parents were informed that he was dying.

Liu Shu and Tian Yuzhao took turns sitting by Edward's bedside in the hospital. One afternoon, Yuzhao pulled a book out from where it had been hidden under his coat. Approved writings—Mao's and Stalin's works and volumes of Marxist propaganda—were the only books allowed during the Cultural Revolution. As Edward had seen, other books—"poisonous weeds"—were burned. As a result, Edward says that he had never been told any stories and had "nothing to give me a dream." He says, "Without stories, I never learned to dream. I did not dream."

However, Edward's father had found a worn Chinese translation of Jules Verne's *L'Ile Mystérieuse,* or *The Mysterious Island,* originally published in China as well as in the rest of the world in three parts. His father had the first, "Shipwrecked in the Air."

Yuzhao sat close to Edward's bed and began:

" 'Are we rising again?' 'No. On the contrary.' 'Are we descending?' 'Worse than that, Captain. We are falling!' "

Even if not for the breathtaking story, the experience was life changing for Edward because his father was with him for the first time in his life. Just as remarkable, it was the first story Edward had ever heard. His condition stabilized as his father read the adventure of the escape from a confederate prison in Richmond, Virginia, by five soldiers who had been fighting for the North in the American Civil War. They escaped in a hot-air balloon, but the journey became perilous when they were blown into a catastrophic storm and the balloon crash-landed on an island, where they struggled to find food and shelter. "It took a month to get through the book," says Edward. The heroic men healed from their journey and miraculously survived. So, too, did Edward. He grew stronger as he vicariously lived life on the island, where the men, reunited and revitalized, prepared a feast of kangaroo soup and suckling pigs. In the middle of the meal, one of the men let out a "cry and an oath." When his friends asked what was wrong, the sailor replied, "I have just broken a tooth!" He drew from his lips "the object which had cost him a grinder," and discovered that it was not a pebble, as they expected. It was a leaden bullet.

His father closed the book. There was no more.

Edward was aghast. He thought, *Someone else must be on the island. But who?* "I didn't have the next volume!" Edward says. "Might these great men be killed? I didn't know. For ten years I didn't know."

Fretting for the safety of the fictional heroes, Edward says, "I lay in the hospital bed and I dreamed. For the first time in my life. I dreamed of many possibilities. Before I had nothing to imagine, but my imagination had been ignited."

China began to dream again, too. The Cultural Revolution ended with the death of Chairman Mao on September 9, 1976. "The reddest sun dropped from the sky of the Middle Kingdom," wrote Anchee Min in *Red Azalea,* her memoir. Mao's funeral went on for days. "Overnight the country became an ocean of white paper flowers. Mourners beat their heads against the doors, on grocery counters and on walls. Devastating grief. The official funeral music was broadcast day and night. It made the air sag." Like Min, who "had no tears, I cupped my face with my hands to hide my face," Edward says, "Everyone cried. I pretended to cry."

The Gang of Four, lead by Jiang Qing, Mao's wife, briefly succeeded him, but they quickly lost power and were subsequently jailed or executed. Only then did the second stage of Edward's life and the beginning of the second stage for modern China start. Edward recalls that the ascension of Deng Xiaoping was celebrated with drums and gongs. Deng "opened our eyes to the outside world," as he describes it. He calls it the period of "wake up." "It was like seeing for the first time after living your life in darkness. I opened my eyes and the eyes of all of China opened to the outside world. For my generation, we saw out of the darkness for the first time. Now I could use my new ability to dream. The Chinese people, for the first time in the life of my generation, dreamed."

Edward, back in Shenyang, attended school in a climate that seemed hopeful. There was news of the world beyond China and foreigners began to visit his country for the first time. "Not one person in China was the same," he says. "Each of us changed. We looked outward."

Until then Edward had been a mediocre student, uninspired by the Mao-centric propaganda disguised as curriculum. He had been turned down for the Little Red Guard because of poor grades. However, after recovering from his illness, Edward took to his studies with passion and seriousness. It paid off when he tested in the top 2 percent in one of the all-country placement examinations. Whereas prior to the Cultural Revolution, entrance was decided on exam results and family background, scores now became everything. Though he earned an average grade for math, he scored extremely high in Chinese literature and won a place at Liaoning University where he was assigned to study biology.

The first week on campus, he went to the university's recently reopened and replenished library. Edward says, "I searched for the second and third parts of *Mysterious Island*. When I found them, I held onto them." He retreated to a corner table in the library and read books two and three, "The Abandoned" and "The Secret of the Island." Ten years of wondering what happened to the castaways was sated during the ten consecutive hours he read. When he emerged from the library near midnight, he was exhausted and his hair stuck up like a mad scientist's. However, walking through the deserted campus back to his dormitory, he felt exhilarated. "I remembered when my father read my first story," he says. "I remembered when I learned to dream."

The biology laboratories at Liaoning had rows of tables with microscopes and dissecting stations. Students strolled the nearby hills with their teachers identifying plants and collecting samples. He met Kong Qin, or Jean Kong, another student in the biology department, on an outing. She was popular, whereas Edward was, he says, "a lonely student." She worked as an editor in a publishing house. While she and Edward had literary discussions, Edward was too shy to ask for a date. "I spent four years dreaming about her," he says.

After graduating, Edward was accepted at the Chinese Academy of Sciences (CAS) in Beijing. He went to a telephone station to tell his parents in Lanzhou. It took five hours to get the call through. "I was so excited when I heard my parents voices five thousand kilometers away," he says. It was his first time using a telephone to make a long distance call.

Edward moved to Beijing and enrolled in first-year classes. Beijing was in the throes of change. When he arrived, students throughout the city wore identification tags. Since his ID advertised that he attended one of China's best schools, he was looked on with admiration. Within a year, however, no one cared. In the emerging China, other things were more valued: the ability to take a taxi, fine clothes, Coca-Cola. While studying, Edward worked as a translator for the CAS in meetings with Western institutions and companies. Sometimes he escorted Western businessmen on their visits through China. Once, he took a group from Germany to the elegant Sheraton Hotel, which had been opened for President Reagan's visit to Beijing in 1987, but he wasn't allowed inside the "foreign exchange currency only" hotel—"not for Chinese."

Jean moved to Dalian, where she edited scientific books for her father's publishing company. Edward called a few times, but a long-distance call cost as much as a week's worth of meals, so they kept in touch by mail. When they met again in Beijing, they went on their first dates. They were married in Beijing in 1987, at a time that China's "opening-up period" was continuing at a thrilling pace and Premier Deng was encouraging China's best students to go abroad to be educated. Of the thirty people in Edward's class, twenty-five ended up in top U.S. universities. Edward's parents advised him to apply to graduate school in the United States. A Texas Tech professor visiting from the United States with a delegation of students also encouraged him. "Come and we will take care of you," he said. Edward took the required examinations and was accepted by Texas Tech and several other universities. The students and professor he had met in China convinced him to come to TT.

At twenty-five years old, Edward left China and arrived in Lubbock, where he answered an advertisement for housing and moved in with an elderly Texan who had never met anyone Chinese. Since it was difficult for the man to remember "Suning," he began calling his boarder "Ed." Later Ed asked to be called by the more dignified "Edward."

The plan was that Jean would come in a year. Edward, with few friends, was extremely lonely. He worked hard on his studies and in the process decided to shift his Ph.D. concentration from biology—"I got tired of counting leafy stem"—to ecology, because of the distressing environmental problems in China, a nation decades behind the West when it comes to research on and policies about pollution. He joined a research project on the renewal properties of fire on grassland that led him to fieldwork that wasn't unlike his parents' research in the Gobi Desert. He was studying the impact of fire on desert soil and plant life by setting fires on the grassland at a site in the desert. He worked during the heat of the day and at night drank beer and talked around a campfire with cowboys who were helping the researchers. They slept under the stars. For his next research project, he worked with a professor who was studying the impact of sewage, shipped by train from New York City, on the Texas grassland. To reach the site, he drove a red pickup truck at 120 miles an hour. "In my cowboy hat, with the windows rolled down in the hundred-plus-degree heat, drinking a root beer, listening to the only radio station I could pick up—an AM country music station—I felt a new type of freedom."

There were several terrifying experiences in Texas, too. Once, three men, shouting racial slurs, beat him up and broke his glasses. But while there were a few isolated examples of overt racism, most Texans went out of their way to make Edward feel at home.

He was amazed by the generosity of some of the people he got to know. In order to get a driver's license, Edward had to take a driver's education course, but he couldn't afford it. A professor paid. When he struggled with some coursework, fellow students and another professor insisted on helping him. A teacher Edward hardly knew volunteered to sign a form that guaranteed payment of Edward's telephone bill so that he could get a telephone in his name. (He had no credit.) "Why does he trust me?" Edward wanted to understand. He says that the supportive environment—the patience, trust, and professionalism of those he encountered—helped his confidence grow. He made some of the closest friendships of his life.

One day Edward wandered into a dark basement of the university and found the deserted computer lab. He had never used a computer, but he was intrigued by the row of Macintoshes. Edward sat down in front of a Mac and pushed a button. Edward would never forget what he saw: the Macintosh's opening icon, a smiling face. Now he says, "Without that smiling face, I may have shut off the computer and never come back. The smiling face welcomed me into a new world."

Intrigued, Edward tried out programs and later connected to the university's Bitnet, an academic network. "I began to think of the Macintosh as my friend, and it led me to other friends," he says. "It was the way I communicated and found out information about my home in China. I played games and wrote letters." It was a revelation to find other Chinese students on the network. "It was my first glimpse of the unifying world of the computer," he says. He explored bulletin boards and eventually started a center for Chinese students in America to gather and exchange messages. With a student he met at Texas Tech, Edward started an Internet organization called Sino Ecologists Club Overseas, which grew to almost three hundred people who knew each other only on the Net. Collaborating online, in early 1990 they published a book about the environmental needs of China. The club evolved into China's first environmental organization, which later moved to the Mainland. It would become one of the most virulent voices speaking out in opposition of China's controversial Three Gorges Dam project.

When Qin arrived in Texas the following February, she, like many Chinese, adopted her homophonic American name, Jean. Besides studying, Jean worked at a Mongolian barbecue restaurant while Edward pursued his Ph.D. throughout the spring. In May, events back in China began to preoccupy them. Edward sat glued each night to the evening news on television in their apartment off the main Texas Tech campus. Most nights the lead report was about an astonishing pro-democracy demonstration that had been growing in Tiananmen Square in the heart of Beijing.

Bo's family can be traced back five hundred years to the Ming Dynasty. His mother, Dong Lihui, was born in Shanghai to a wealthy family from Ningbo. The family lost its fortune during the Japanese invasion of 1939. Their home in Shanghai was burned to the ground and the family fled to Hangzhou, where within a year Lihui's father, mother, and brother died of tuberculosis. Relatives raised her and her surviving sister. Bo's father, Feng Zhijun, is the great-grandson of an imperial doctor who served in the Forbidden City until the 1911 Republican Revolution. Zhijun's father was the chief editor of Reuters News Service in Beijing until the Japanese invasion. He refused to work during the occupation, so his second wife, Bo's grandmother, supported the family by opening a public bath and sauna. A few years later the family moved to Shanghai. Zhijun and Lihui met in high school in Shanghai and married soon after Zhijun graduated from Shanghai University of Railway Engineering in 1962. Both became teachers.

When the Cultural Revolution began, Zhijun was denounced in one of the first purges of intellectuals. "Our father was a cocky professor who refused to renounce his beliefs," says Tao Feng, Bo's elder brother. "Being a professor was bad. Being a cocky professor was a disaster." Zhijun was sent to a reeducation camp where he worked in the fields and fished. "If you think you're so smart, start shoveling pig shit," was a well-known Maoist saying. Lihui was also forced to stop teaching, but she was allowed to stay in Shanghai because she was ill. Regardless of her health, however, she had to work as a menial laborer in a lightbulb factory, earning the equivalent of $2.30 a month.

Tao was sent to live with Zhijun since Lihui was too ill to care for him. He remembers his father's suffering and humiliation. It was typical of the time; intellectuals, including the best teachers, were denounced in meetings at which they were forced to kneel for hours, sometimes with dunce caps. Many were beaten. The torture was psychological as well as physical. Chinese Muslims like Feng Zhijun were forced to tend pigs; writers were forced to burn their books. Tao remembers watching Zhijun being hung precariously outside a multi-storied building where he was forced to write slogans denouncing the capitalist class and celebrating the proletarian revolution. As the son of Feng Zhijun, Tao was told to lick the soles of the boots of children of Red Guard officers.

Bo was born two and a half years after Tao in 1969 near Si Pin Lu, the pentagon, or five-pointed square, of Shanghai. He stayed with his mother, who cared for him as best she could. She relied on the help of neighbors and relatives, but when there was no one available, she tied Bo to his crib when she had to leave to work.

In the early 1970s, Zhijun and Tao were allowed to leave the labor camp for a rural village, where Zhijun read the works of Mao Zedong and taught "Mao Zedong thought" to illiterate peasants.

Zhijun and Tao were allowed to visit Lihui and Bo in Shanghai more frequently and finally, in 1976, the Cultural Revolution ended and they returned home. The family moved into a small apartment with cracked walls and no heat on the grounds of the Railway University. They shared one kitchen with four other families. They bathed in a wooden bucket filled with water boiled on a communal stove. Zhijun and Lihui began teaching again.

Traditional education was stopped throughout the Cultural Revolution, but the brothers were sent to the nearby elementary school in 1977. It had a vast roof of blue tiles and peaked eaves and a courtyard for basketball and soccer. Films were shown on Saturday evenings. From Hong Kong and even some American movies, they were Bo's first glimpses of a world beyond China. Sitting in the common room on folding wooden chairs, he was riveted to Clint Eastwood and other western heroes. He eagerly anticipated the Saturday night movies and remembers that once, when he earned a bad score on a math test, his grandmother forbid him from a Saturday night showing of *Hamlet*.

Zhijun left his teaching position to found a new university, Shanghai's Institute of the Science of Sciences, a school of management science. He also wrote the first of more than forty books about the modernization of China, management, and philosophy. He entered politics in the 1980s and secured a ministry-level legislative position as a member of the Standing Committee of the People's Congress. He also worked his way up in the Democratic League, a liberal minority party of intellectuals. Though opposition parties had little power in China, they played a significant role in the cultural and academic development of the nation after the Cultural Revolution.

At eleven, Bo was sent to a boarding school, the Shanghai Muslim School, where he lived for six years. One of the best schools in the area, it cost his parents the equivalent of $3 a month for room and board, in addition to the semester fee of $5. Eight students lived in a room. For three years, Bo's bunkmate was Shen Baojun, with whom Bo broke windows and nearly got kicked out of school for their practical jokes. In fact, Bo rarely studied and should have been expelled, but he was allowed to stay in school because Zhijun, though like Bo a terrible student, was one of the Muslim School's most prominent graduates. Whereas Tao was a model student, Zhijun made excuses for Bo. "My father told people that I will change. He always defended me," says Bo. "People got on my case, but not him."

On most weekends, Bo rode forty minutes by bike to see his parents at their new government-owned flat in Shanghai when Zhijun began working for the provincial government. Tao came home from Fudan University High School on weekends, too. Bo was seventeen when Zhijun became a vice chairman of the Democratic League. Three months later, Bo graduated. He didn't get into college because his scores were abysmal. He tried but failed to get into hotel school. His family viewed it as his last chance for a reasonably successful life. His parents were frustrated with Bo's aimlessness. He was excited by ideas but overwhelmed. He had no sense of what he could or should do next.

In 1985, Zhijun watched the backlash to the Cultural Revolution, when Western ideals seeped into China and brands such as Nike and Marlboro became status symbols. Serious culture has already suffered, Zhijun said, instructing Tao and Bo to work hard for China and "keep your culture" even as he prepared to send his sons abroad. Since Mao had dismantled China's system of higher education, the best education for Chinese students of their generation was in the West, and Zhijun sent Tao, after graduating from the prestigious Harbin University, to graduate school at the University of Toronto.

Zhijun had visited the United States with a delegation of foreign dignitaries after the Cultural Revolution as a representative of the city of Shanghai. He went to Washington, D.C., after the Watergate scandal and was astounded to see how the American system held its leaders accountable not to the power structure but to the people. After the visit, a friend Zhijun made in the United States referred a colleague named Agnes Wang to him when she visited Beijing in 1987. Wang, a psychologist practicing in Marin County, California, became a close friend.

Upon her departure from China, she offered to host Zhijun's youngest son if he came to the United States. Zhijun and Lihui viewed it as a potentially lifesaving opportunity for Bo. Later in the year, she received a letter from Zhijun confirming that Bo would come. There were no other options for Bo in China, so Zhijun expressed his blind hope that his son would find himself in the United States. Without high academic marks, it was almost impossible to get a visa to leave China. However, Wang's sponsorship allowed Zhijun to send Bo, then eighteen, to San Francisco. He arrived on October 23, 1987.

Wang met him at the airport and drove him over the Golden Gate Bridge to the rural Marin County, California, town of San Geronimo. Her redwood home was set amid horse ranches and grazing holsteins. She set him up in his own quiet bedroom—the first time in his life he had his own room—on the second floor above her psychotherapy office. Bo says that he sat alone in the room and felt he had made a terrible mistake: "To say that my world was upside down is inaccurate. It was beyond any word you can use to try to describe it."

Knowing little English, Bo could speak to few people other than Wang. There were no buses to San Francisco, where he could have visited the Chinese neighborhoods, so he was isolated in Marin. He enrolled in the English as a Second Language program at the College of Marin. Arriving for the first day of school, he wore his only suit, a pressed white shirt, and black brogues. Everyone else was in T-shirts and jeans; Bo was embarrassed and confused.

He missed China but noticed and became excited by the open and relaxed atmosphere in America. His English quickly improved and he made friends with people he met at school and at his job at a local Chinese restaurant.

After a year, when Bo turned nineteen, Agnes gave him a birthday party. She invited her family and his few friends from school and his job. It was the first birthday party of his life. In China, such a luxury was unthinkable. Agnes says, "He seemed almost overwhelmed by how much his senses could be fed by tenderness, softness, and open communication."

Agnes moved to a small house in nearby San Anselmo, another sleepy and woodsy town where Bo worked eleven-hour shifts as a busboy at the Mandarin Garden Restaurant for $15 a night. He got other jobs at Chinese and Japanese restaurants, making his way from one restaurant kitchen to another, cleaning fish, cutting fish, steaming rice, setting tables, clearing tables, taking orders, and carrying away and washing dirty dishes.

When he wasn't in school or working, he hung out at cafés, where he struggled to read the *New York Times* and eventually struck up conversations with patrons and the staff. He learned enough English to pass the required exams and began taking general classes. Courses in art, particularly photography and film, were a revelation to him: "They taught me a language of expression that I didn't know existed." His photographs from that time depict isolation. One shows a darkly lit wrought-iron bench that is beautifully composed but haunting. The bench is empty. He had an exhibition of his photographic collages. The theme: searching.

Bo trained as a waiter and worked at the Robata Grill, a sushi bar and traditional Japanese grill restaurant in Mill Valley. He used his wages to fund his first eight-millimeter movie projects. As his English improved, he read and talked Sartre, Camus, and his favorite author, Martin Heiddegger, at San Anselmo's Café Nuevo, what he calls "the existential hangout."

Bo missed his family, but he was filled up with images, ideas, the ability to create art, and his new friends. China seemed like a distant dream until the spring of 1989.

CHAPTER 3

WHAT IS HAPPENING TO OUR COUNTRY?

Tiananmen Square is the symbolic as well as physical heart of Beijing. A hundred acres large, the square begins in the north at the Gate of Heavenly Peace, the Tiananmen, which is the entryway, guarded by a pair of stone lions, to the Forbidden City, the stronghold of a long line of Chinese emperors ("the oracle from which the emperor exercised the mandate of heaven"). Beyond the gate one sees the golden fish-scale tiles of the Forbidden City's peaked roofs.

To the south of the massive square is the Mao Zedong Memorial Hall, where Chairman Mao's body lies in state. Catty-corner from the gate and across the street is the Great Hall of the People, with massive columns, its facade crowned with the nation's symbol: a crimson seal with five golden stars encircled by a golden wreath. The large star represents the Communist Party, the four smaller ones the classes in Chinese society—the peasantry, the workers, the capitalists, and the bourgeoisie.

On April 15, 1989, former Communist Party general secretary Hu Yaobang, an elder statesman sympathetic to the nation's reform-minded university students, died. Chinese citizens flocked to Tiananmen Square to mourn Hu. Four days later, a hundred thousand students gathered on the broad steps of the Great Hall of the People, where the commemoration of Hu evolved into a protest for far-reaching reforms. On the following day, three student representatives climbed the steps of the Great Hall and demanded to meet Premier Li Peng. Kneeling, they awaited a response, but none came. Instead, the students were denounced in an editorial in the party's newspaper, the *People's Daily*.

Students throughout China were incensed. At forty universities they boycotted their classes and, on April 27, hundreds of thousands of students marched through Beijing to Tiananmen Square. The mass of students, wearing headbands and waving red flags, grew over the days that followed. On May 4 their representative read a declaration that called upon the government to accelerate political and economic reform, guarantee constitutional freedoms, fight corruption, and allow a free and private press. In mid-May, several students in the square began a hunger strike and elected student representatives began formal talks with the government, but they broke down. Twelve of China's most respected writers and scholars presented an emergency appeal that called on the government to acknowledge the student movement as "a patriotic democracy movement" and asked the students to end their hunger strike, but neither the government nor the students responded. The crowd at Tiananmen Square grew still larger. Similar protests sprung up in cities throughout China.

On May 15, the third day of the hunger strike, Soviet leader Mikhail Gorbachev arrived in Beijing for the first Sino-Soviet summit since 1959. The government had planned a welcoming ceremony in Tiananmen Square, but it was canceled. Meanwhile, the large contingent of foreign reporters in Beijing for the summit began to cover little else but the students. The world was now watching.

On May 18, the sixth day of the hunger strike, Premier Li summoned several student leaders for a televised talk at the Great Hall, but his condescending attitude further alienated and angered the demonstrators, who next learned that the government prepared to declare martial law. They called off the hunger strike and began a mass sit-in. Premier Li gave a speech that foreboded what was to come: He called for "firm and resolute measures to end the turmoil swiftly."

Martial law was declared on the following day, though the army's advance toward the city was blocked by swarms of students and their supporters. On May 23, the troops pulled back to the outskirts of Beijing. According to *The Tiananmen Papers*, based on controversial documents smuggled out of China and published in 2001, there was dissension among the leaders of the Communist Party about how next to respond. The moderate factions advocated a conciliatory posture, but the reigning "elders," led by the retired but still in control Deng Xiaoping, made the decision to use force.

On May 27, a student group called the Alliance to Protect the Constitution unanimously voted to recommend that the students end their occupation of the square, but the resolution was rejected by a rival student group. (This group was lead by Chai Ling, the student leader who gave a famous interview in which she predicted bloodshed.) The demonstrators unveiled a ten-meter-high "Goddess of Democracy," replica of the Statue of Liberty, on May 30, and another hunger strike began on June 2. On the third, Premier Li, reportedly following Deng's instructions, issued the order for the troops to reclaim Tiananmen Square "at all cost." At ten o'clock at night, the army opened fire. Students and bystanders were shot. Tanks and armored personnel carriers moved toward the center of the city and more shots rang out, killing and wounding more citizens. The following day, June 4, at around one in the morning, troops surrounded Tiananmen Square and awaited further orders until four o'clock, when the men on the second hunger strike negotiated with military leaders to allow the students to leave the square. At five, students, along with their teachers and supporters, fled at gunpoint.

In Texas and California, Edward and Bo watched it all on television. Bo was stunned by the demonstrations and devastated by the violence. The images played and replayed on CNN: a bloodied innocent bystander, crumpled on the pavement; a lone protester, startlingly brave and dignified, standing up to five tanks, stopping them; People's Liberation Army soldiers beating and dragging students away; a line of personnel carriers snaking ominously toward the square. It staggered Bo and others who revered him that Deng ordered the attack; the man who had moved China forward, opening it, embracing foreign education and the entrepreneurial spirit, betrayed the people of China. When protests in support of the students broke out around the world, Bo wrapped black ribbons—shrouds—around trees in the town of San Anselmo. "It was a day of blackness," he says.

Bo tried but was unable to reach his parents on the phone. In China, Zhijun, who by then had moved to Beijing and had become a member of the People's Congress, supported a restrained response to the students even as the army moved in. Bo spoke to Tao, who was studying in Canada, sharing their worries about Zhijun. After several days, Agnes Wang was awakened in the middle of the night by a telephone call. A man's voice said that he was calling from the city of Xi'an. Was Bo available? When Bo picked up the receiver, the man said, "Your father wants you to know that he is all right," and then the telephone went dead. Later Bo learned that the mysterious call was from an uncle who feared calling the United States but had been instructed by Zhijun to let his boys know that he was safe. However, while safe, Zhijun's sympathy for the students at Tiananmen Square proved costly. He was reprimanded and his political ascent—there had been reports that he would rise in the government higher than any non–Communist Party member—was stopped. In July, Zhijun had a serious heart attack. When he recovered, he went back to work as a respected professor, writing influential books, and he remained a representative for the Ningxia region in the People's Congress and became the vice chairman of the Central Committee of the Democratic League. However, there was no more speculation about his political ascent to the highest echelon of power in China.

In the aftermath of the massacre, thousands of people in China were arrested and hundreds fled. In the United States, many of the tens of thousands of students who were studying abroad, known as *liu xue sheng* (literally "study abroad students"), resolved never to return. In the United States alone, forty thousand *liu xue sheng* were granted green cards that permitted them to stay and work in America.

It was exactly what China didn't need. The nation's recent history is one that has undermined its progress by discouraging (or worse: punishing, beating, even killing) its best and brightest. Since the 1950s and throughout the Cultural Revolution, millions of Chinese intellectuals were humiliated, tortured, or murdered. Immediately after the Cultural Revolution, many of the relatively few intellectuals who returned from the West suffered subtle or overt persecution. Finally, after the cessation of the purges of intellectuals following the Cultural Revolution, Deng sought to remedy the nation's brain drain by sending large numbers of China's younger generation abroad to be educated in the West. The idea was that they would return to rebuild China. However, the June 4 massacre disaffected many Chinese students and they decided to forsake China forever. The consequences were dire. China once again pushed away its intellectuals. Who would lead China, whether its government or its industry, into the future?

Of the millions of students of Bo and Edward's generation who had gone abroad through the 1970s and 1980s, only a small fraction returned. The number dropped to almost none after June 1989. Chinese students in the United States typically graduated and got jobs in American companies or academic institutions. Indeed, the Chinese students who remained in the United States were one of the most successful groups of immigrants ever. Chai Ling, the student leader at Tiananmen Square, escaped China for the United States and eventually founded an Internet company in Boston.

After the deaths and the retreat by the students, the Tians, like Bo, were dumbstruck; "in a fog of the shock," as Edward puts it. "Nothing would ever feel quite the same after that," he says. Friends and teachers at Texas Tech sent flowers of condolence and called to offer their support.

The uprising was squelched and the news cameras turned away from China, but both Edward and Bo spent weeks and months that grew to more than a year attempting to figure out what they should do. Edward says, "We thought and thought after that: How could we sacrifice to help the people of our country? It became the central purpose in our life."

With his discovery of art, Bo was, for the first time in his life, a dedicated and enthusiastic student. He became obsessed with film. Truffaut, Buñuel, and Godard became his gods. He watched *Blow-Up* more than a hundred times. He thought he might try to become a film-maker. However, as he says, "that was before June 4." "Afterward, I saw that China didn't need three-hour art films, and I felt that I could not stay away from the challenge of helping the people of China," he says. "Our country was in turmoil, but what could I do?"

Edward contemplated abandoning his education. "Pure science? Ecology? I felt that a Chinese person did not have that luxury. I had to do something else, something that would have an impact. All I thought about was what I could do."

Things quieted down in China after the government tanks had withdrawn and the square was silent once again. In the United States, however, both men felt something similar: They had to return to China. Bo explains, "How could we turn away from China? Our families? The Chinese people?" In China, he had been aimless. His wife Heidi says, "He would have been a crushed person there." But he couldn't relate to the traditional path for Chinese students in the United States—that is, college as a means to a well-paying job. Art did intrigue him, but he was, in his words, "a lost person." No longer. "It's better to light a candle than curse the darkness" goes another Chinese saying. Bo had a purpose.

In the early 2000s, Bo and Edward would be followed by a mass return to China—what would be described as a "reverse brain drain." Well-educated, highly motivated, and extremely talented Chinese graduates would begin to return to the Mainland in droves. In the early 1990s, however, it was a radical decision. A friend told Bo, "China is the old world. China is death. How can you go back?" Bo responded, "I have no choice."

Edward became a voracious reader. He loved biographies. One he read at the time was the story of Steve Jobs and Apple Computer. Edward had a special affection for Jobs's Apple Computer because of his experience with the Macintosh computer's smiling face. Reading the biography, Edward was inspired by Jobs's vision of the power of the entrepreneur to change a society. In addition, he began to think about Jobs's idea that computers are not just computing machines but tools with the inherent ability to change lives. "Steve Jobs gave the computer industry a much greater goal: to make a better world," Edward says. "The idea began to consume me."

While Jean helped supplement their scholarships by working at the restaurant, Edward translated books, including Lee Iacocca's biography, for Jean's father's publishing company. He earned $1,000, an astronomical figure for him. In spite of his doubts, partly because he didn't know what else to do and partly because he felt he would be in a stronger position if he continued, he completed his Ph.D. in 1992 as he continued to ponder his next step.

In 1991, Edward returned to China for a visit. Deng had recently made a tour of the southern region to reinvigorate economic reforms, and Edward was amazed with what he saw during a train trip from Shenzhen to Guangzhou. "Construction was under way everywhere along the railway track. Men who had been farmers and peasants were working all night long to build the nation. There was a desperation to change their lives now that there was the opportunity." He wrote Jean. In his letter, he said that he wanted to be part of the shaping of the new China. He was still unsure how until, the following year, he discovered "the answer."

A friend in Texas convinced Edward to come to hear a speech by Senator Al Gore, who was running for vice president of the United States on a ticket with Bill Clinton in 1992. Gore's speech stirred Edward. He knew that Gore's father had passed the interstate superhighway bill. Now the son was talking about a different type of highway—an information superhighway. Edward had used the Texas Tech network, but Gore described the Net in a way Edward never considered. He said that the information superhighway, if a national policy, could transform a nation. How? By elevating communication, knowledge, and education—exactly what China desperately needed. "Information! Access! That's what we never had!" Edward says. "Information threatens the status quo. We were starved for information when we grew up. The Internet could give Chinese children access. It could connect them."

In Chinese, Internet is *huo liang wang*, combining the characters for "connect," "each other," and "network." Edward thought, From an isolated province, each Chinese person could connect. They could visit the Louvre or any library in the world. In addition, as an ecologist Edward responded to Gore's point that the Internet is clean technology. He thought, If China continues to grow based on oil there will be a global disaster. China needed to progress but not only by industrialization. Instead, moving ideas and information, whether by e-mail, videoconferencing, or the Web, was an alternative to more cars, buses, factories, and trains.

The famous dissident Wei Jingsheng, after emerging from fifteen years of solitary confinement in a Chinese jail, once said, "People long for change, but they despair of it, so they go into business." Edward turned to business, too, but it wasn't because he despaired of change. The opposite. Edward is a vitally clear, methodical thinker, and he weighed the possibilities. China, with a fifth of the world's population, was stuck in the past, while the rest of the world raced toward a new, as yet undefined, future. The overthrow of its system might seem like the answer to some, but Russia's post-Communist decline showed that it wasn't enough. Besides, the Tiananmen Square massacre proved that protests were not enough to change China. So how? Edward thought it through. What did China lack? Opportunity, abundance, openness of information and communication, a strong educational system, and a strong economy to benefit the people. "I began to see that there was another way to change China and I thought that it might be effective," he says. "It was a radical idea at the time, but I became devoted to it. Like Steve Jobs said, entrepreneurship could be an engine of a society's progress. Deng had brought in the possibility of entrepreneurship to China, so that door was open. Jobs also described how computers could empower individuals. Gore's vision held that the Internet could connect them. The combination could be the force that would transform a nation." He read more, audited management and computer classes, and contemplated his next step.

When his father read him Jules Verne, Edward learned to dream. Now the confluence of Tiananmen Square, Steve Jobs's idea of entrepreneurism and Al Gore's vision of the Internet lead to a "big dream"— a dream for China and a direction for his life. Edward would work to connect China to the information superhighway. The mix of these transforming experiences and influences—ones that lead him in a clear, passionate direction—propelled Edward forward and made the decisions that followed remarkably easy given what was at stake. Edward from then on had an unbridled determination to change history, unconcerned about obstacles or the odds against him.

Now the thinking seems normal: The Net will change the world. Most of us still understand the revolution inherent in the Internet even in the sobered, post-cybermad fin de siècle of the late nineties. However, when Edward set his sights on the Net in 1993, it was anything but certain that it would survive. The Net and the World Wide Web were way far-out, chancy propositions. To show how improbable they were, consider that this was before the official founding of either Netscape or Yahoo. At the time, few people had modems, and those that did slogged along at twenty-four hundred or fewer baud. E-mail was beginning to take off, but most people still sent their mail by the U.S. Post Office or, when there was a rush, by Federal Express or fax. Netscape cofounder Marc Andreessen had written his now-famous browser, Mosaic, but, as he once told me in an interview, "The work was fun, but no one was taking it seriously." At that time most people considered the Internet "a ploy—a low-bandwidth ploy at that—nerds and scientists and typing," Andreessen said. "That's what everyone thought—Microsoft and everybody else. I thought, 'I may as well work on this now, and then when I get out of college I can go work for Silicon Graphics or Time Warner or TCI.'" In the United States, it was a pretty far-out idea, but in China? In 1993 it bordered on lunacy.

Edward foresaw the potential of the Internet, however, and he wrote to his family, friends, and colleagues about the way that the technology could radically transform China. "Inherent in the technology is the most invaluable resource in a free world: a free flow of information," he said. "In the past, a few people controlled information in China. Before the printing press, books—the teachings of Confucius, anything else—could only be read by the elite. The printing press meant the possibility of the equal distribution of information, but it was controllable. The elite could block information that was presented in a physical form. Not in a digital form, however. Because of this technology, China, too, will be the open world."

His revelation lead to the third stage in his life and in modern China. "We are at the dawn of the age of the information-technology entrepreneur," he says. "Information technology has the power to bring a renaissance to China." Here's how: "The driver of entrepreneurs are dreams," he says. " 'Why do you want to start a company?' 'What is your fundamental motivation?' A dream. With the Internet, young children will be able to download beautiful stories and begin to have their own dreams. Their dreams will lead them to become entrepreneurs. When they do, they will become engines of change. China will slowly evolve."

Edward was committed to return to China, but there was a problem. Jean was adamant in her refusal to come along. Jean wanted to raise their daughter Stephanie in the West, which now felt like home. "At the time, American schools had teachers who were dream makers, not dream breakers," says Edward. "I knew that part of my job was to make China a society of dream makers, too." Edward rationalized leaving them behind. "I couldn't only think about my own child," he says ruefully. "I also had to think of all of the children of China. They need to have the opportunity that Stephanie has."

They decided that Jean would stay in Texas and Edward would commute. The arrangement would inevitably mean long separations. Indeed, he would miss Stephanie's first words, her first steps. "What choice did I have?" Edward asks. "A Chinese poet writes, 'Because of the irresistible calling of history, you have no choice.' China has struggled for two-and-a-half centuries. My generation has that burden on our shoulders."

To prepare for his return, Edward immersed himself in the world of technology and business, further educating himself about the Internet and analyzing ways he might work on the technology in China. He wrote the first article about the information superhighway for a Chinese academic journal, and from the environmental bulletin board he spun off a discussion forum devoted to the promise of information technology in China. The first subscriber was a UCLA student named James Ding. Without meeting, James and Edward realized that they were kindred spirits with similar political, environmental, and social concerns, including an interest in technology and entrepreneurism.

Ding's Chinese given name is Jian; he called himself James when he came to America to study in 1988. An exceedingly bright computer science student at the Institute of Science and Technology Information in Beijing, James was encouraged by a friend to apply to UCLA, where his experience with PCs got him a job in the university's computer lab as a systems administrator. IBM had donated network cards and systems software and Novell donated operating systems to the university, but no one knew how to use them. James studied the manuals and installed the first LAN (local area network) at UCLA in 1989. He was working on the network during the spring and early summer of Tiananmen Square.

Before graduating in 1990, James completed his job of connecting the campus network to the Internet. Next he was offered the job of systems analyst at the University of Texas in Dallas, where he met Edward in person. The two discussed ideas for a partnership until early 1993, when Edward finally approached James with a proposal for a company called AsiaInfo Daily News, a spin-off of his bulletin board. The plan: to sell subscriptions to a news service that focused on China. They agreed to try it. Up and running by the end of the year, AsiaInfo Daily News included translated political, entertainment, and financial news from and about China. James handled the technical side and Edward the management, marketing, and sales, which were slim. The two may have been the first entrepreneurs to learn that people don't pay for content on the Net.

Edward never lost sight of his goal of returning to China, though he hadn't figured out how to make the leap across the ocean. With James, he strategized, "How could we get this technology home?" The two made an exploratory trip to China. The first transglobal network in China had been developed for a joint project between Beijing University and the Stanford Linear Accelerator in the mid-1980s. Scientists from both institutions set up a network that was up and running in 1985, and the first e-mails were shot back and forth between Palo Alto and Beijing. When Edward and James visited in 1993, government-owned China Telecom was exploring the Net not as a commercial platform, but for academia. "It was a toy they wanted to play with," James says. He and Edward also learned that a handful of Chinese ministries and state-owned businesses, watching the development of the Net in the West, were beginning to research networks for their departments and enterprises. Back in Texas, Edward and Jean moved from Lubbock to Dallas, where he and James founded Business Development International, officially BDI. The company's plan: to wire China. What that meant, however, was inconceivable to them.

CHAPTER 4

CHINESE MIND

When I met him in 1992, Bo was dating a dear friend, Heidi Van Horn. They met at the Robata Grill, in Mill Valley, California, where they both worked, she while studying for a degree in social welfare at UC, Berkeley. Bo had a mixed reputation at the restaurant. He was regularly fired and rehired. One time a customer insulted a waitress and then stiffed her. Bo followed the man outside and let loose with a barrage of insults. His manager took Bo aside and said, "I admire your chivalry, but you're fired."

Heidi is exceedingly smart and poised, and has a sharp and subtle wit. She has an evenness and good-heartedness that you can count on. (Bo does.) Meeting Heidi, who is delicate and radiant with dark-brown eyes and hair cut sharply at her shoulders, was life changing for Bo. "When I arrived at eighteen, I felt as if I landed fifteen thousand feet underwater. Years of struggle—paddling, paddling, paddling—but when I met Heidi I finally came up for air," he says. He was determined to return to China, but with Heidi and her family, Bo felt more grounded in the United States.

After they dated for two months, Heidi and Bo became engaged and moved in together with her parents. In China, it is more typical than not to live with one's parents or in-laws, so it was a logical arrangement for Bo. Heidi's parents immediately fell in love with Bo, who cooked for them and stayed up late with them talking politics and business. By living with Heidi's parents, they were able to save some money. Early in 1993, Heidi encouraged Bo to return to China. He hadn't seen his mother and father in five years. Their savings weren't enough to pay for the trip, so Bo sold his prized possession, a white VW Beetle, a gift from a second cousin living in Los Angeles.

En route to China, they stopped in Japan, where they stayed with Bo's boyhood friend Shen Baojun, who was living and working in Tokyo. When they were alone, Baojun slipped Bo an envelope with $2,000 in cash and presented Bo with a state-of-the-art video camera. "Pursue your dreams," said Baojun. When Bo pressed him about the money, Baojun confided that he earned the money—and more, *lots* more—running a gambling parlor for the Yakuza, the Japanese Mafia.

Bo's parents were eager to see their son and his fiancée, who they heard about from Tao. They knew that she was well educated and came from a close family that had warmly embraced Bo. Indeed, when Heidi and Bo arrived in the middle of the harsh Beijing winter, Zhijun showered Heidi with gifts and Lihui prepared festive meals. They communicated with sign language and a few English words. It was a struggle to converse, particularly since Bo was an impatient translator. (Zhijun would address Heidi for fifteen or more minutes, but Bo's translation lasted a line or two.)

Once, when Heidi went to the outdoor market with Lihui, Bo sat face-to-face with his father for the first time in five years and discussed his life. Bo says that he will never forget the conversation. Zhijun counseled Bo to remain in the United States so that he could develop his skills and, as he gently put it, "find your calling." He said, "Your opportunities to do that are abroad. China needs the inspiration of the ideas from places that have not been isolated and stymied." Bo told Zhijun about his frustration after Tiananmen Square and his commitment to return to China to contribute in a meaningful way, but Zhijun said, "I have complete faith in you. You will come back when you are ready."

Bo took Heidi to visit Hangzhou. Arriving by train at Hangzhou Station, they took a taxi around the city's beautiful lake, with its thick layer of floating Jurassic irises and luxuriously weeping willow trees. The cab dropped them at the Xinxin Hotel, but without a marriage certificate they weren't allowed in. It was a vestige of the Cultural Revolution, when hotels and guesthouses required an official letter allowing in a guest, and unmarried couples would automatically be forbidden. Since they couldn't afford two rooms, there was no choice but to head back to Beijing. The next train didn't leave for three hours, so Bo hired a rickshaw and they circled the lake under the raining green willows. They talked and watched the fat stars and the swinging red lanterns on the boats that navigated the dark water.

In Beijing, they took a breathtaking gondola ride to the top of the magisterial Great Wall. They went on a full-day walk through the Forbidden City, one of the rare reminders of ancient China that survived the rampages of the Cultural Revolution. (The Forbidden City survived because Premier Zhou Enlai sent the army to protect it from the marauding Red Guard.) Bo showed Heidi Shanghai, where, introduced by a friend of Zhijun's, Bo met with a businessman from Taiwan who was interested in developing real estate in the Pudong District. Bo was offered a three-month job working for the developer, who needed help navigating Bo's hometown—identifying sites and negotiating deals.

Bo had never been interested in business. He says that his family was somewhat contemptuous of it. However, he viewed the job as a way to make some real money and to understand more about the changes in China since he departed five years earlier. Working, talking to everyone he met, and observing the opening culture, Bo says, "I realized what needed to be done: China needed to rebuild. I didn't understand my role, but I saw the very basic needs of China: solid buildings, good water, dependable phone lines, roadways." In spite of the new camera from Baojun, he felt that he was ready to leave behind his art for good. "Coming back to China completed the destruction of my search for art," he says.

After three months, as fall arrived and plum and cherry trees bloomed along the main streets of Beijing, Heidi returned to Berkeley to graduate from Cal and begin a new position at the university's Public Service Center. Bo followed soon after. They were married in April.

The wedding was held at Heidi's mother's new home on a grassy hilltop with a view of the small town of Penngrove in Sonoma County. Heidi and Bo invited dozens of friends, and her entire family was there along with numerous dogs with plastic cones around their necks to prevent them from gnawing at their stitches. (Her stepfather is a veterinary surgeon.) Bo's guest list included Tao, who flew in from Canada, where he was completing his Ph.D. in applied statistics, and Agnes Wang. His parents sent a faxed greeting from China; they couldn't afford the trip.

Heidi's father, a stockbroker, introduced Bo to a businessman who was interested in promoting Visa cards in China. (Credit card use in China was then extremely rare, though Visa began to make inroads in the country in the mid-1990s.) In the course of an informal meeting, Bo was offered the job of educating his associates about China, translating, making introductions, and helping to negotiate deals and review contracts. It was unexpected, since he had no training in business other than his brief stint in Shanghai, but Bo is a quick study and did what came naturally. The advice Bo gave was based on simple logic rooted in his understanding of China. He began reading business books, too. "Business is logic," Bo says. "Logic and instinct." The last thing he expected was to like business, but, he says, "I fell in love at first sight."

Through Bo's bosses at the Visa job, he met Sandy Robertson, a legend in the Silicon Valley, in late 1994. Robertson was cofounder of the venerable investment banking firm Robertson Stephens. The company had financed numerous high-tech and biotech start-ups, including Sun Microsystems, Applied Materials, Dell Computer, National Semiconductor, Ascend, Compuware, Intuit, Pixar, Excite, and AOL— transactions worth more than $73.5 billion. "Robertson has helped put together more venture capitalists and more deals than any other investment banker," writes Udayan Gupta in *Done Deals*. "[He] typifies the entrepreneurial spirit that bankers have had to possess to do business with venture capitalists and venture-backed companies." The time of the meeting with Bo was fortuitous. "It was too soon to go to China," he explains. "But not too soon to be thinking about China. We weren't ready to set up a big China operation, but it was time to start investigating."

Robertson says he was looking for "the right guy" in business schools when Bo came along. They met in the Bank of America building twenty-six floors above San Francisco's financial district. Robertson, in his banker's wool suit and a monogrammed shirt with a red tie, sat at a large desk in front of a wall of photos of his family and a snapshot of he and his wife, Jeanne, escorting President Clinton down the steps of their home. Lucite tombstones—bankers' trophies from their deals—packed three bookshelves. It would be tough to name a technology company that wasn't represented.

Robertson, with silver hair, piercing eyes, and an avuncular lilt in his voice, succumbed to Bo's passion, charisma, and prodigious intelligence. "It was instant," he says. "Bo had that thing—an innate people sense, an incisive mind, and remarkable leadership skills. It's all the stuff they try to but can't teach in business schools. He didn't know our business but I thought he could get it all right."

Robertson asked him if he would come on board to help Robertson Stephens make its first inroads into China. To complete the deal, Bo had to meet the firm's partners, which he did on the following Friday.

Bo arrived at Robertson Stephens in a three-button blue suit bought for the occasion and was led to a seat at the end of an oval conference room. A dozen of the company's partners sat around a mahogany table. After small talk, one of them cut to the chase. He turned to Bo and asked, "Why are you useful to us?"

Bo, his arms unfolded and his palms up on the banker's table, answered carefully. "I will help you learn how we see China," he said. "Many people look at China blindly or they see one part and think it is the whole thing." He sat upright, locking his arms across his chest. Using the analogy of the blind man who thinks he knows an elephant because he knows one part—the tusk, the trunk, the leg—Bo said, "China is much more than any single piece. Westerners who fail to see the complexity of China will fail in China.

"There is a lot of excitement about the Chinese market," he continued. "However, it is much more complex than selling toothbrushes to one point three billion people. China is in the middle of a reform—of ownership, social and economic values, and the structure of our society. After wars and revolutions, the Chinese people now have a chance to build our country, to create a social and economic foundation for everyone in China." Succinctly explaining the enormity of the challenge, Bo said, "There is a high risk to investing in China. There are barriers, from language to regulations to the difference in the personality and values of Chinese compared to Americans. However, I can help you to bridge those barriers and can create exciting opportunities. The Chinese language is about the description around things, around an experience, and it reflects the Chinese people, who view life more obliquely than Westerners, less didactically. The Chinese language and culture are not direct. So I must bring the West into the more poetic, symbolic world of China, but also must bring Chinese business into the direct, exact world of the West." He told the group that Chinese culture is like old Chinese coins, round on the outside with a square center. "We have our center, our principals, our way of being, and we have our way of appearing, which is not always the same as what is inside."

Bo was hired. His job: director of the China Group—initially Bo, an analyst, and a secretary. "It was trial by fire," Sandy Robertson says. "Now that he was an investment banker, Bo had to figure out exactly what that meant. We had to train Bo quickly." Over a series of dinners and in meetings at Robertson's office, his new mentor gave Bo a crash course in business—"business 101," Robertson says. Robertson gave Bo a reading list so that he would learn the basics about technology and instructed Bo on the history of the information-technology industry. The promise and the potential. Moore's law. The impact of Netscape. The transforming potential of IT on people's lives. Bo immersed himself in the Robertson Stephens research, talked to the company's analysts. Bo listened, took notes, and asked questions. "He got it, got it fast," Robertson says. "Bo knew how to dig things out. You have to do that—be scrappy, think on your feet. He does."

When Robertson spoke about the ways that technology had and would change people's lives, Bo's mind always went to China. "I began to put the pieces together—to see how IT could help China leap forward." That was it. A *real* leap forward was possible. China had been the world's preeminent nation in the agricultural economy, but when that economy gave way to the industrial age, the fundamental rules changed. China stagnated. The massive structural advantages China had built up in the agricultural economy became irrelevant. Poor countries in Europe that just emerged from the Dark Ages leapt into the industrial age; the West caught up to China in fewer than a hundred years. Bo saw how a new age levels the playing field. A new age is upon us—the information age and the knowledge economy—and the playing field is once again leveled. China is in the position of the European nations a hundred years ago. Now was an opportunity—a rare historic opportunity—to leapfrog ahead. Even after the Cultural Revolution, China had some of the essential components that would be required to do so. Since the nation discouraged the liberal arts—studying them was potentially dangerous—China had educated a generation of scientists and technologists. Bo saw how capital could create entrepreneurial environments for them—computer scientists, managers, and engineers could learn how to create and build companies. He saw how inefficient state-owned enterprises could give way to fast-moving start-ups of the type that created the Silicon Valley—with Sandy's help. "If we capture the wave," Bo said, "we have a chance to become a great power again—to build as great an economy as anyone else does over the course of the next fifty to one hundred years."

Bo knew that the name Sandy Robertson would open doors and lend a certain amount of credibility, but the success or failure of the China Group was entirely up to him. He was entering waters that were not only uncharted by him, but anyone. That is, everything in China was new and being reorganized. Business and politics in China used be done exclusively through something called *guanxi wang,* the crucial social structure in China based on relationships and position. It ran deeper than anything the Silicon Valley has ever known. However, the old ways of doing business can now work against you. Says Internet consultant Duncan Clark: "You can be completely *guanxied* up and fail miserably. In fact, *guanxi* can be a hindrance because you are tied to inefficient ways of working. Now it's all about whether or not you can get the job done." It meant that there was no one to look to as a role model, no one to call on for advice when it came to entering China itself. He took the job more seriously than anything in his life. Besides his growing conviction about business, he felt a personal responsibility to Robertson, who had given him this chance. Sandy's stories about the great American entrepreneurs weren't lost on Bo. He thought, Maybe in all of China there is one out there who will found a great new company that will make a difference. Maybe that person will bring technology to China that can begin to change lives. Maybe I can find him.

Bo, who headed back to China with a Robertson Stephens business card and not much more, began a search for entrepreneurs working on promising technologies. Robertson says, "When he left, we had no preconceived ideas about what he would, or should, find. Something that works here might not work there. Bo had to figure it out himself. After all the talk, there was no road map. He set out and did what I would have done: He looked for the greatest people he could find."

Bo had heard about Wang Zhidong, the youthful founder of SRS, the company behind the nation's most used homegrown software package, and Yan Yanchou, the middle-aged software designer who wrote the original Chinese DOS. "They were legends in Zhongguancun," Bo says. "They were known to be the smartest of the software geniuses."

He searched them out at a decrepit school building set on a deserted street at the outskirts of Zhongguancun. To get in the school building, Bo climbed over the metal fence, pushed open a brittle door, and wandered through dimly lit hallways, crunching broken glass.

Wang Zhidong, with haywire hair and frameless glasses that sat cockeyed on his nose, looked up from a computer monitor. He wore a casual shirt and sandals and was, at twenty-five years old, a boyish man who sat tightly hunched in his chair, a bona fide geek with pencils in his breast pocket. Zhidong and Bo were, within minutes, drinking tea and talking, not about business but about history and technology. Later, Zhidong told me that he is distrustful of businessmen but responded to Bo immediately. "He wasn't like the others," Zhidong said. "Here was someone with whom I could talk—about business, about life."

When Bo, a storyteller by nature and perhaps even by trade, first told me about Zhidong, he had his friend growing up on a duck farm in rural China. It has become a sort of urban legend, or at least a legend in the Wang story. When, in 2000, *Forbes* included Zhidong on its list of the wealthiest Chinese entrepreneurs, it noted the duck farm background.

The point is sound even if the facts aren't. Zhidong's grandparents were rice farmers outside the treeless, rural village of Humen in Guangdong Province near where the Pearl River spills into the South China Sea. Humen has a notable history. On June 3, 1839, Lin Zexu, an imperial envoy, drew the line against the opium trade in Humen. Over the course of twenty now-famous days, he ordered the digging of two pools, each forty-five square meters, added salt water, and poured in twenty thousand cartons of confiscated opium. After stirring the opium-filled water to be sure the drug had dissolved, he opened the pools' floodgates and let it drain into the sea. Britain, which had profited enormously from the opium trade (and therefore the rampant addiction that was crippling China), attacked China, and the Opium War raged for the next three years.

Zhidong's parents left the farm and moved to Humen proper, where they became teachers. They were poor—one of the poorest families in the poor village. Six people lived in a small room. In spite of the family's poverty, the fertile land meant that there were no shortages of food during the earlier Communist-era famine or the Cultural Revolution.

Though they were intellectuals with college degrees, his parents, Wang Peisen and Chen Bangsong, were, Zhidong says, "little people in a little town," so they avoided the type of persecution that befell Bo's and Edward's parents. Instead, they worked in their school throughout the Cultural Revolution, although they were forced to change the lessons they taught from academic course work to propaganda: the recitation of articles by Mao and quotations from *The Little Red Book*.

Zhidong was born in the first year of the Cultural Revolution, 1967. Literally meaning "Rising Sun," the characters of his name promised the glorious future of China under Mao. He attended Hong Qi, or "Red Flag," an elementary school, where Zhidong became a Little Red Guard and recited Mao's praises. Bo didn't invent the duck farm out of thin air. One summer, when he wasn't in school, Zhidong traveled to a middle province to the duck farm owned by his relatives, where he worked in the fields and tended the fowl.

The end of the Cultural Revolution, followed by the downfall of the Gang of Four and subsequent establishment of Deng Xiaoping as Communist Party general secretary, changed Zhidong's prospects.

Education was Deng's highest priority, and there was a sense of urgency about it that came from the need to make up for a decade of lost time. Zhidong's parents once again taught their pupils in the range of general academic subjects, and Zhidong, who was ten years old, began to receive a solid and rigorous education. He attended high school in the neighboring town of Dongguan, where there was a strong incentive to do well; each semester that he ranked in the top three of his class, the 50 yuan tuition was waived. For him, "studying hard was a way out of the life presented to me," and his parents never had to pay the fee.

In high school, Zhidong discovered electrical engineering, working obsessively in the school's wireless lab, dismantling and reassembling radio receivers and telegraph machines. While Bo and Edward left China to continue their education, it wasn't an option for Zhidong. However, his score in the top 1 percent in the nation on the national college placement test lead to a full scholarship, which included 27 yuan a month for food, in the prestigious Wireless Telecommunications Department at Beijing University. In the summer of 1984, he boarded a train and rode to the capital city, where he settled into his new home, a Beijing University dormitory with eight roommates.

Zhidong had read about computers and was fascinated by them, but he had never seen one. On a tour of the campus, he discovered the computer lab, where he lingered so long that the laboratory head asked him if he knew how to use the machines. Zhidong blurted, "I do." When he was escorted to a computer and couldn't turn it on, he covered his fib by saying, "This is a different one."

Freshmen who wanted to use the lab had to take an introductory course in computers unless they could test out by taking an examination. After studying for a week, Zhidong got the highest score of any student. He soon transferred from electrical engineering to computer science and virtually took up residence in the computer lab. It wasn't uncommon for the teacher to find him in the morning asleep in the lab, his head resting on a keyboard. "Computers gave me the opportunity to express myself in ways that I never could," he says. "It was almost a way of realizing things—making real things that you can imagine. Computers give individuals power. You can enter all of your thoughts into the computer and the computer will follow your instructions exactly. It will produce the result you desire."

The lab was both Zhidong's academic and social life; he didn't date and his friends were "geeks and nerds"—his description—like himself. In 1987, Zhidong's junior year, he began to earn money as a freelance software engineer in several nearby Zhongguancun companies. When he graduated, a teacher sent him to apply for a job in the government's Agricultural Ministry at a factory that produced yogurt, where he was hired to help the facility automate. A year later, he returned to the Zhongguancun, where he worked as a freelancer at dozens of companies in every imaginable department, from customer support to business development to accounting and sales.

Just before graduating, Zhidong met an entering student named Liu Bing. "As I was walking out of the university, she was walking in," he says. After enrolling, she asked a teacher to introduce her to someone who could teach her about computers. The teacher asked Zhidong, who agreed, particularly when he learned that Bing was studying English literature. He wanted to learn English and he thought they might teach each other. "But she didn't learn computers and I didn't learn English," he says. With Bing, a smart and confident woman with a bob and almond-colored eyes, he talked about history, their mutual interest, while she pursued her degree and he worked at numerous technology companies.

In 1989, Zhidong was recruited by the university's Computer Science Center, where he went to work at the Beijing Founders Group, a commercial spin-off company that made computers and software. He was working there in 1989 during the demonstrations at Tiananmen Square: "I was looking at my computer screen instead of at the streets." Now he says, "I was interested in a different kind of revolution. I knew that the technology would have a great impact on China if our people had access, too."

At the Founders Group he worked on a desktop publishing software package used for typesetting and new versions of Chinese DOS, the program written originally by the famed computer scientist Yan Yanchou. Next he was responsible for creating a developing platform for Windows with a group named BD Win, for Bei Da Win, after the Chinese name of Beijing University. (*Bei da* also can mean "stupid egg," which is a running joke on the university campus.)

In 1991, while contemplating a job offer in Singapore, he proposed to Liu Bing.

"I would like us to be married before I leave the country," he said.

They married secretly in a Beijing courthouse because it was against the regulations for a student to marry before graduating, and Bing had a year to go. After the rushed wedding, Zhidong turned down the job abroad. Now she continues to accuse him of tricking her into marriage.

Next Zhidong left the Founders Group for Suntendy Electronic Information Technology and Research Company, where he marketed Chinese DOS. With a classmate and a friend he soon spun off his own company called Chinese Star, which developed the first Chinese-language adaptation of the Microsoft Windows operating system. His partners were interested in diversifying the business into restaurants and real estate, but Zhidong was only interested in computers, so they parted ways in 1993. When Wang left C-Star, IBM, Microsoft, and a number of state-owned enterprises offered him jobs. "He couldn't just take a job," sighs Liu Bing. "He wanted to do something important for China. He is very patriotic, believing that we must work to build China and help the Chinese people." Bing says that she tried to discourage him. "I wanted him to live a calm life with regular vacations, but Zhidong didn't listen."

He met Yan Yanchou at a computer conference in 1986. Yanchou, who had studied at the Junior High School for Workers, Peasants, and Soldiers in Hunan and then at the National Research Institute, knew more about computers than practically anyone in China. Yan's first experience with a computer had been in 1979, when a teacher asked him if he was interested in designing a computer "motherboard." He was given four months. "I did it because my salary was extremely low," he says. "If I can do better, I can promote my wage." He taught himself, working with a storeroom of components. His board became one of the first homegrown microcomputers in China.

In 1983, the government arranged for the first two IBM PCs to be delivered to China. They came in pieces because of a government regulation that stipulated that computers couldn't be imported, and so they had to be assembled in China. One of the PCs was sent to a factory, where engineers worked on cloning parts. The other was given to Yan's institute. "It is very difficult to develop Chinese software," he says. "When the government decided to do it, no other institutes were willing to take the task. So the computer ended up on my desk."

Yan hid away for fifteen days reading documentation and in four months designed a breakthrough program called Chinese Character DOS, or CC-DOS, which was crucial to the development of both a PC industry and a PC consciousness in China. "The most difficult obstacle was to make English software receive Chinese characters," Yanchou says. "Also, displaying Chinese characters is different. English characters appear in twenty-five rows. For Chinese, there are ten rows because the characters are more complex. How to make a virtual display monitor was our big problem, especially since the PC was such a weak platform." Even so, Yanchou's PC became the first to run Chinese-language software.

With this and other 6800 series computers, Yanchou taught himself more advanced hardware design and programming. His next assignments were the creation of the first and second computers used in industry in China—a payroll system for a private company and a computer that controlled a flour mill. In 1984, IBM began shipping in large numbers of PCs in an agreement with what was then the Ministry of Electronics. Yanchou created the first word processor so that users could type Chinese characters. There were two ways to do so. One system required the typing of character parts, but it was extremely complex, since one had to remember thousands of them. (There are 140,000 characters in the Chinese lexicon.) The alternative way to type Chinese characters, called pinyin, was much easier. Computer keys are assigned sounds or syllables and users make words by combining them. If a sound is represented by more than one Chinese character, a pop-up menu of choices appears. "Now keystrokes are like brush strokes," he says.

The IBM was too limited for many uses, particularly when it was working in Chinese, so Yanchou suggested souping it up. To do so, he was sent by his institute to work at the newly formed National Computer Bureau. Throughout 1984 and 1985, he made a CRT controller and display card, and he imported all Chinese characters into it. He made a fortune for the National Computer Bureau, which held the patents on Yan's inventions.

Yanchou finally left for a private company "to go to swim in the sea," as he puts it, in 1986. "There was no national guarantee, so there was risk, but I was interested in the experiment." He worked for an electronics firm writing application software for power plants. In the middle of the year, Yanchou was in charge of a conference for Microsoft's Windows 3.0 in Beijing, where he met Wang Zhidong. Soon they were discussing a new company that would create software for Chinese PCs, though it would take a year for them to go forward.

When they did, Zhidong and Yanchou were turned down by a long list of investors before the Stone Group, a Hong Kong–based importer of typewriters and Compaq PCs, agreed to provide seed capital. Stone Rich Site (SRS) set up shop in the abandoned school building, where it was joined by Zhidong's younger brother, Wang Zhigang, another brilliant engineer who, like Zhidong, had graduated from Beijing University. Rounding out the team was another engineer and Liu Bing, who left her outside job to manage the office.

The Stone funding, enough for little more than salaries and rent, was running out and Zhidong needed more money if SRS was going to grow. He met with bankers, but nothing happened until Bo showed up.

When Bo arrived at the SRS office, the newly minted banker explained that he could help Zhidong expand and meet the growing tide of foreign competition by raising a substantial round of funding. Zhidong listened carefully. He was a computer genius without business experience, but he could analyze the scenarios Bo presented. From Sandy Robertson, Bo understood that entrepreneurial companies usually need more from their investors than cash. They also require advice, resources, contacts, and sometimes hands-on management assistance. Bo promised all of that, offering to help Zhidong in any way that he could. Zhidong was impressed. After several more meetings, he agreed in principle to work with Bo.

Bo told Zhidong that he needed a business plan to show to potential investors, so Zhidong wrote one. When he proudly presented it to Bo, however, Bo quickly read it and handed it back to him. "Try again," he said. "This time, no adjectives."

It took eight months, through mid-1995, to finalize a thorough plan. When it was completed, Zhidong fell into a panic. He worried that investors would take over his company and told Bo that he was backing out of their deal. Bo was flabbergasted. What if, after everything, Zhidong backed out? He could be fired from Robertson

Stephens if he didn't have anything to show for all the time he had spent on the SRS deal.

When Zhidong came around and agreed to go forward, Bo felt it would be useful to get Sandy's take on SRS as well as other companies he was investigating, and he convinced his boss to come to China. Before Sandy arrived, Bo was a nervous wreck, afraid that Sandy would discover what Bo, at his worst moments, felt: that his new hire was incompetent. Bo thought, My boss has no idea of my real position in China, which is worthless. But Sandy, experienced and respected, must know what he's doing. If Sandy says I can do it, if he sees something in me, something I don't see, maybe he is right. If Sandy says I can do this, maybe I can.

Bo rented a large Buick so he could drive Sandy around Beijing. He introduced Robertson to the leaders of China's established technology companies, including hardware and software companies such as Legend, the Stone Group, and the Beijing Founders Group. He brought him to meet Yan Yixun, the vice chair of the Academy of Science, who was a family friend and an influential supporter of the technology industry in China. He also introduced Sandy to Zhongguancun. Robertson found it enthralling, "somewhat primitive, operating on a shoestring, but with great energy and lots of good ideas," he says. "They were doing impressive things with pretty shoddy equipment. You'd walk up some dingy stairway and enter a room and there would be software designers working at a feverish pace on some breakthrough invention. It was like going back to the sixties in the Valley. You'd walk into some dingy lab and meet some kid named Gordon Moore." (Moore, the inventor of the microprocessor and developer of Moore's law, was the cofounder of Intel.)

After he introduced them in the Friendship Hotel lobby, Sandy and Zhidong talked at a small cocktail table. Sandy confided one of his misgivings about investing in the Internet in China. "It's not really possible to get anything much online, is it?" he asked. "The government blocks access to everything from the West, right?"

Zhidong didn't immediately answer. He instead reached over and picked up his suitcase and slipped out a laptop, which he set up on the cocktail table and fired up. While the computer booted, Zhidong lifted a telephone on a nearby coffee table and removed the phone cord, plugging it into the back of his computer. When the PC was ready, he configured his system to place a call. In a minute there was the familiar static of modems talking and then, the connection made, Zhidong did some quick clicking on the keyboard. He turned the computer to face Sandy, who smiled when he saw the Robertson Stephens website.

"I see," he said. "I guess we have the wrong impression."

By the end of the meeting, Sandy was relating war stories of deals with the big-name U.S. tech companies. He and Zhidong parted with a firm handshake and the promise that they would meet again when Zhidong next visited San Francisco.

After Sandy agreed that SRS was worthy of his company's name, Bo took the business plan and met with potential investors throughout Asia and the United States. The reactions at first were grim. The worst was when a U.S. investor, after hearing the company's plan, asked, "What happens if there is another Cultural Revolution? Do you have a contingency plan?"

It took dozens of meetings, but Bo pulled off his first deal as an investment banker, raising $7 million. Nowadays $7 million seems like milk money to most technology companies in Silicon Valley, but at the time, Zhidong says, "it was more money than we could imagine." The deal closed at a time when SRS had two hundred employees and a somewhat less grim Zhongguancun office near Beijing University. Inside a series of cramped rooms, employees worked at quartets of cubicles on metal desks with swinging lamps. In what was once a cafeteria, a line of engineers worked at folding tables placed end to end with workstations plugged into a chain of extension cords. Yan Yanchou remembers, "It was a very dangerous place. Each time a new employee arrived and plugged in a computer, we would take a deep breath and wait to see if the whole company would go down." What it lacked in flash, however, SRS had inside its computer hard drives and on its network: the work of a growing number of China's best computer scientists.

Three years before Microsoft released its Chinese version of Windows, Zhidong and Yanchou released a program they designed called RichWin, which translated standard English Windows to Chinese, allowing it to read and display Chinese characters. RichWin became essential on most Chinese PCs. Not only did it translate and adapt Windows, but it translated and adapted non-Chinese Windows software, which meant that Chinese speakers could use most foreign programs, from Microsoft Word to Quicken. In addition, the program had Chinese fonts, instant translations back and forth between Chinese and English, and a Chinese-English dictionary. Even users of the late-comer to the Chinese market, Microsoft's Chinese version of Windows, used RichWin, because of the resources and additional functions. Zhidong explains, "It's not just a simple translation, not just a language issue. Importing programs from the West is less useful for Chinese. They miss the complexity of the way we think about problems." The difference between Zhidong's software and Windows is that it is "at its nature Chinese, reflecting the Chinese mind," rather than a foreign structure imported and translated. Yan Yanchou explains that "we had a long-term vision: We wished to add more Chinese culture into software." Even the building blocks of Windows are Western. English characters are single byte; Chinese are double byte. In addition, Chinese users require a range of utilities that are unnecessary in the West. The inherent difference is why Zhidong believed then (and continues to believe now) that homegrown software will prevail in China.

Since Bo's deal, SRS has continued to grow quickly. By 1997, 90 percent of new PCs sold in China came bundled with Zhidong and Yanchou's program. RichWin became the biggest-selling software program in China. About half of the 8 million copies in circulation in 1999 were paid for. Zhidong wasn't overly concerned that the rest were boot-leg copies in a country rife with counterfeiting, because his program was cheap enough for many people, people who could afford computers in the first place, to pay for. Their motivation: authentic, bug-free programs plus documentation and support.

Zhidong's success didn't go unnoticed by Microsoft. Early in the summer, Steve Ballmer, Bill Gates's partner and the Microsoft president, paid a visit to SRS. Ballmer knew that Zhidong's company had the potential to undermine Microsoft's position in what would likely become the world's largest software market. First, RichWin could drastically cut into sales of Chinese Windows. (At the time, most computers in China came loaded with standard English Windows; without RichWin, Chinese users would have to purchase the Chinese version, too.) In addition, Windows could theoretically become extraneous and expendable underneath RichWin much like DOS was expendable underneath Windows 98.

Ballmer, who returned twice to SRS, argued that there was synergy between his company and Zhidong's. His goal seemed to be to build a relationship that would ensure that Zhidong would continue to create programs that were not only compatible but also dependent on Windows. For the short term, there was no other choice, since Windows ran on almost every PC in China. However, that could change. It was possible to imagine a future generation of RichWin as an operating system for PCs or Internet terminals in place of Windows. A better program at a cheaper price could inspire Chinese computer makers, and even foreign makers who wanted to sell in China, to make systems based on RichWin. It's why the first of a series of magazine articles appeared that asked the question: Will Wang Zhidong become the Bill Gates of China?

CHAPTER 5

IF YOU BUILD IT

There would never have been a place for the Internet in Maoist China, but Deng Xiaoping unintentionally paved the way. It began when he brought capitalism into China in the late 1970s, upping the ante with his famous call to arms in 1992. "To get rich is glorious," he said, turning the country's Marxist-Maoist agenda on its head. Since then, as Orville Schell observed in *Mandate of Heaven,* "China is as drugged on business as it was on politics."

The first wave of capitalism in China resulted in the privatization of state-owned enterprises and thousands of Sino-American, Sino-Asian, and Sino-European joint ventures, many involving the partial extrication of state-owned enterprises from labyrinthine, inefficient bureaucracies run by the government. By 2001, SOEs accounted for 28 percent of the country's gross industrial output, down from 55 percent in 1990. Besides the China-only joint ventures, there were newly founded companies that opened the KFC franchise on the southwest corner of Tiananmen Square, made Chinese-German Volkswagen Santana cars, and sold Western medicines—eventually, even Viagra—in pharmacies throughout the country. The government continues to struggle with privatization. In 1999, it announced that six thousand companies established and run by the military, from industrial conglomerates to karaoke bars, were to become independent. In early 2000, thousands more government companies privatized or spun off independently run enterprises with the government as a shareholder.

The opening of the country also meant that myriad foreign companies giddily arrived in China with plans to sell their products—everything imaginable—to the largest populace on Earth. Though pre-WTO China was risky, as Robert Allen, then the CEO of AT&T, noted, "next to China, all other opportunities around the world pale." It sometimes took more complex and dicey joint ventures with Chinese companies, but the multinationals set up shop in China.

Then came the entrepreneurs. There were thousands and then tens of thousands of them. Neither restructured government enterprises nor branches of foreign companies nor traditional joint ventures, both homegrown and foreign-born businesses sprouted in the fertile post-Deng landscape. Countless of them. They tried everything imaginable: copy centers, trendy magazines, instant noodles, bicycle-messenger services, Western-inspired brew pubs. Some were independent, and some were founded with government investors or partners.

Edward Tian and James Ding realized that a joint venture was the only realistic way they could set up in China, so James proposed a company to colleagues at his former institute, the Institute of Science and Technology Information. When it was approved, they finally had a firm foothold on the Mainland. Next they set out to round out their initial team. Soon there were ten engineers working in small rented offices near the University of Texas in Dallas and in a four-story walk-up on a Zhongguancun side street. To help them manage the company, they hired Liu Yadong, a student from the Mainland who had earned a degree in nuclear science and neuroscience at the University of

Maryland and who then became an expert in handwriting-recognition software. He moved to Mountain View, California, with a company he cofounded that was acquired by Adobe. There he worked on Adobe Acrobat until he joined Edward and James.

The first potential customer called in 1995. It was China's first big-board stock exchange in Shenzhen. Its managers wanted to automate their operations with what would be the nation's first Web-based information system for stock trading. Edward and James met the company's technical team and introduced their company. James presented a plan for the technology that he would employ based on his experience at UCLA and UT. When they were asked how much the job would cost, James and Edward looked at one another. They hadn't thought about that part.

That night, in their hotel room, they came up with $2.2 million, which they presented the next day. When the manager said, "Fine. Let's start," Edward and James looked at one another. Had they grossly underbid?

The man said, "How's forty percent up front? Where should I wire the money?"

Edward sheepishly said that BDI didn't yet have a bank account because the joint venture wasn't officially registered in China.

"Edward!" snapped James, trying not to loose his composure in front of their first client.

Edward admitted that he paid $9,000 to register the venture in China, but the "expert" he hired to put through the paperwork had disappeared with their money. No, they weren't registered.

"We have an American bank account for AsiaInfo," he said, perking up. "How about that?"

The man said no, the stock exchange could only use a domestic contractor, so Edward borrowed a telephone and made some calls. A friend in Beijing agreed to allow BDI to use his registration number and bank account.

When they left the company's office, James and Edward hopped into another taxi. James seemed elated, but Edward was too traumatized to celebrate. The next day, however, when they returned to Beijing, Edward received a telephone call from the friend who had provided the bank account. When he hung up the phone, he walked to

James's desk. With a restrained grin, he said, "We just got funds wired to us. Eight hundred and eighty thousand dollars!"

"Eight hundred and eighty thousand?"

James was speechless.

Setting up temporary shop in Shenzhen, a vibrant and modern city across the waterway from Hong Kong, James, Edward, Yadong, and their staff got to work. The job went relatively easily because of James's immaculate design. Edward flew back and forth between Shenzhen and Dallas. Besides visits to Jean, he bought the hardware, mostly equipment made by Sun and Cisco. For software, James and his team modified shareware programs. The system was running in six months. Their first customer called to say that the company was extremely satisfied with the system, which compared to the one in current use by the New York Stock Exchange. It was easy to use, "logical," and created great efficiencies in the operation.

BDI moved into a warehouselike office in Zhongguancun, where the partners bartered with the landlord. In exchange for wiring the building, the rent was free. Edward next heard from a representative of Sprint International, an early Western entrant into China. The Ministry of Information Industries, which oversees China Telecom, had contracted with Sprint to begin to build a dial-up-based commercial Internet in China. The reason the government agency called a foreign company was that no major player in China had the expertise to build it.

At the time, Sprint was spreading itself thin in the United States while building its network, and the company didn't have enough high-level engineers to tackle the Chinese contract. That's why Sprint decided to subcontract the job to BDI. Edward, James, and their team hired more engineers and built, with $4 million worth of imported computers and routers, the first phase of an Internet system in China, a link between Shanghai and Beijing, and connected both cities to the international Net via leased undersea telephone cables. When that leg was complete, Sprint hired BDI to add a node in Guangzhou, and by mid-1994 the three cities could connect to one another and the global Net over telephone lines. Testing them, James sent the first e-mail message over the commercial Internet to a friend in the States. He also connected to the World Wide Web and read the news from the United States.

In 1995, China Telecom prepared to extend the contract with Sprint for the next phase of the project, the wiring of additional Chinese cities to the network it named ChinaNet. Sprint by then no longer wanted a subcontractor and planned to cut BDI out of the deal. When Edward heard about Sprint's decision, he told James, "Let's get this contract ourselves."

"How?" James asked. "We're nobodies. They are Sprint!"

Edward responded, "We're going to get the deal."

Renamed AsiaInfo and newly independent (now that it had a foothold in China, Edward and James incorporated in Delaware and dissolved the joint venture), the company successfully won the contract. They had a track record, and the government preferred working with a domestic company if possible. More business came AsiaInfo's way from provincial telecommunications companies. Edward and James made strong presentations, convincing company representatives of the importance of getting wired. It was a patriotic duty, Edward would say. "Only by preparing with the world's state-of-the-art technology will China be able to compete in the information age," he said. "Yes, it is a small business at this time. Nevertheless, it will grow. It will be as important in China as the railway network. More. Don't be late to join the information age."

After reading books about management from the West, Edward tried to be a new type of boss in China, inspiring rather than intimidating his employees in order to get them to work long hours for relatively little pay but immeasurable amounts of glory. The speeches he is now known for may have been less polished then, but they were just as impassioned. "We are bringing the information superhighway to China!" he declared. "You can tell your children and your grandchildren that you brought China into the modern world, that you made our nation great. You can tell them that you changed the lives of our people. What will this technology do for China? Change the way we think, the way we view the world. It will lead to a better life for all of our people."

James explained the technology to prospective customers, Edward sealed deals, and Yadong, who became chief operating officer, negotiated the terms. Once contracts were signed, James's team of technicians set out to implement the technology. James himself solved the most challenging technical problems. There are many examples. While the team was installing a system at a provincial telephone company, a mainframe computer crashed. Two of the company's staff scientists who had built the system were unable to fix it. James asked if he could try. The system was up in an afternoon. Another time, he successfully wrote a program for a Cisco router in forty-eight hours because AsiaInfo couldn't wait the weeks, perhaps months, that it would have taken for the software to be delivered. When the local Cisco rep inspected the program, he was flabbergasted, noting that it seamlessly replaced a program that a team of software engineers in the Silicon Valley had created over the course of a year.

Following the model of America Online, the founders of Shanghai Online, a venture of the Shanghai Post and Telecom Administration Bureau, prepared to launch what would be the first Chinese portal. In mid-1995, Yadong heard from the head of the project, who asked if AsiaInfo was interested in bidding on the contract to build and manage the company's website, which included both the development of Shanghai's local Internet backbone and the supporting software for Shanghai Online's information services. Other bidders included IBM, Digital, and Sun Microsystems. Next, MCI called to see if AsiaInfo wanted to bid on a potentially enormous deal.

For the MCI meeting, James and Edward flew to Washington from Beijing and Dallas, respectively. Edward's discount airline flight stopped in Memphis and Detroit en route to the capital. By the time he arrived, James had rented a car and the two sped to MCI, where they met with the team in charge of the Asia project. The meeting went well; AsiaInfo wound up with a lucrative contract.

When Edward and James checked into their hotel, a message from Yadong—marked urgent—was waiting. If AsiaInfo wanted a shot at the Shanghai Online deal, James had to be at a technology meeting in Shanghai in forty-eight hours. James called the airport to find the next flight out. To catch it, they had to race. James sped on the freeway toward Dulles International. However, when they were almost there, James shrieked. "We're at the wrong airport! My flight leaves from National!" He cut across three lanes of traffic and careened off the freeway. He drove to National, arriving twenty minutes before his flight. He parked in front of the terminal, jumped out, grabbing his single carry-on bag, and threw Edward the keys. "I'll call you from Shanghai," he said.

Edward's flight back to Texas was to leave in two hours from Dulles. The problem was that he had never driven in Washington, and it took him six hours to navigate the freeways to the other airport. He missed his flight. To use his discounted ticket, he had to wait another two hours for the next flight to Memphis. When he arrived, he had to wait three more hours for a flight to Miami. From there, he waited two hours for the Dallas flight. He collapsed on the couch when he got home, but Jean greeted him with the news that Yadong had called to say that he had to immediately leave for Shanghai, too. The Shanghai Online deal hung in the balance. Frayed, aching, and nearly hallucinating from exhaustion, Edward nearly cried in the taxi back to the airport. The next flight out took him from Dallas to Los Angeles and then on to Tokyo, where he hopped a Beijing flight. In Beijing, he caught the last shuttle to Shanghai.

Edward was desperate to sleep when he arrived at the hotel. However, when he went to Yadong's hotel room, an extraordinary sight greeted him. Yadong, James, and fourteen engineers were working on computers lined up on a series of card tables. They had dragged the beds out of adjoining hotel rooms to construct the makeshift office. A soundless TV played as the engineers busily wrote sections of the new presentation, calculated technical specs, and designed the remaining pieces of the online system. While James caught Edward up to speed, the mother of one of the engineers arrived with a platter of fried noodles, which she had delivered by rickshaw.

In the morning, the complete AsiaInfo team assembled in a crowded Shanghai Online conference room. The two founders, operating on a combined total of a half dozen hours of sleep over the previous seventy-two hours, launched into a presentation. Edward set the stage. "IBM? IBM is a stalwart, trusted name," he began. "Is that what you need? IBM? Digital Electronics? Sun? Big and respected corporations? Or do you need an innovator?"

The largest companies in the world got that way because they were the innovators of their time, Edward said. However, he continued, "they don't understand the Internet. We have great respect for IBM, but IBM in China is just another vendor. We have great respect for Digital, but in China the company is just another vendor. Sun? A reliable vendor in China. With AsiaInfo, however, you choose a company that uses the appropriate vendor to create a solution designed not for everyone but for you. We will be devoted to your technology needs. You will be one of thousands of customers for the others. For us, you will be our most important focus."

Edward told the Shanghai Online executives about the failure of the biggest Western corporations, including IBM, TCI, and Microsoft, to create a digital platform for future information technology and communications. "They were invested in old technologies," he said, sipping from a glass of water. "They didn't understand the Internet. They dismissed it. Meanwhile, the Net grew and a small company that no one had heard of, Netscape Communications, created an industry based on the technology. A start-up defined the future. Why? Because the Internet is the fastest-moving technology history has ever seen. On the Internet, innovation, not brand names, is the watchword."

He handed the presentation over to James, who detailed the system they had tailor-made for the company. By the time he wound down, they sensed that they had won the $3 million contract.

Securing the contract was easy compared to creating the system, even with the elaborate blueprint. The toughest challenge came when the AsiaInfo team received an untested piece of hardware from 3Com, an enormously complex AB8000 access server, which no one in China had seen. The machine wouldn't communicate with the other components in the system, even after countless attempts by the crack AsiaInfo engineers. Meanwhile, the Shanghai Online bosses were pressuring Edward for a demonstration of the system. In spite of tireless work, the engineers were stuck. In tears, one gave up, saying it was impossible. James worked consecutively for ninety hours. On the fourth morning, he showed up at the temporary AsiaInfo office in Shanghai and plunked down on Edward's desk with a self-satisfied if demented look on his face. When James told him that he had done it, Edward, noting his friend's glazed dark eyes, untended werewolf hair, and grimy clothes that hadn't been changed in four days, responded, "You need a shower and a new T-shirt."

AsiaInfo designed and implemented systems management and application software modules for accounting, billing, travel, and games, plus intelligent full-text search engines in both English and Chinese. The system has since been adopted as the industry standard by many other online content providers in China. If not a Chinese AOL, Shanghai Online went on to become a leading Internet content provider. AsiaInfo continued to work with the company to maintain and upgrade its systems while tackling commissions for new customers such as the telephone company in Guangdong Province, where AsiaInfo had sealed its reputation because of its successful wiring of the Shenzhen Stock Exchange. By the time it completed GuangdongNet, AsiaInfo had forty-six employees. However, the Internet was growing slowly in China. After all of the company's hard work, the number of Chinese Internet users could be counted in the low thousands. Edward worried, What if he built the network and no one came? At his most optimistic moments, he felt it was only a matter of time before the numbers would grow. However, what if he was wrong? What if competing technologies pushed the Internet aside? What if the Internet in China remained obscure and elitist?

Brooding about this and in a particularly grumpy mood, Edward arrived at the Holiday Inn in Beijing for the first AsiaInfo Christmas party at the end of 1995. He slumped past the bar, where Americans and Chinese were sipping beer under some hastily strung Christmas lights. The lobby newsstand had just been stocked with the new issues of Western magazines. One caught his eye. He picked it up and stared at the cover. The magazine proclaimed it the year of the Internet. He flipped to the article. There was the story of Netscape, which went public in the West and woke America and the rest of the world to the fact that the Internet had arrived, ushering in a new era. The article told the story of the Net pioneers, who the magazine called "the people building the technology that could unite the world so geography no longer matters." Edward thought, I'm not alone in this. Maybe it will be all right.

The AsiaInfo staff were in a private room with a bizarre Western-and-Chinese-style buffet that included dumplings, stinky tofu, and a roast turkey with cranberry sauce. Edward entered the party with a broad smile and hearty greetings to his staff, the magazine tucked underneath his arm.

In mid-1996, AsiaInfo had $2 million in the bank when Edward, who continually monitored the U.S. market, decided that the company should refocus its priorities. Now that most of the nation's Internet backbone was complete, he felt that AsiaInfo should concentrate on the next stage of Internet development, which he termed "Infomation." Taking his lead from Netscape, he decided that AsiaInfo should work on private data networks for Chinese industrial companies. As he explained to his team, "The industrial age brought automation; the information age will bring 'infomation,' when companies will automate all of their internal operations."

James and Yadong begrudgingly supported Edward when he reorganized the company, putting a third of the staff on his "infomation technologies" initiative. However, after three months the division had only signed one customer, Heilongjiang Telecom, and the deal was a disaster because the company's employees rebelled and refused to use the complex system. A month later, the managers of the infomation project came to Edward and begged him to pull the plug. The head of the sales unit dragged Edward to a café, where he threatened to quit if Edward didn't rethink the operation. When Edward refused, the man resigned on the spot.

By Christmas 1996, a third of AsiaInfo's employees, almost all of them in the new division, quit. James, Edward, and Yadong fought. Still, Edward said they had to press on—given time his idea would work.

In a quiet voice, Yadong explained that they didn't have the time. He showed Edward the company's ledger, which indicated that the infomation initiative had drained most of their $2 million savings. There was no choice. Edward had to agree to end the initiative, which he finally did in January. "It was the biggest humiliation to be so wrong," he says.

To regroup, Edward organized a retreat. A friend recommended the "healing setting" of picturesque Monterey, California, three hours south of San Francisco on the Pacific coast. Edward, James, Yadong, and the company's deputy chief engineer, Michael Zhao, met there in late January. Edward began by apologizing. "I was too emotional," he said. "It's not the way to make decisions."

James and the others milked the groveling. At last James said that they would never forgive Edward and they should go get some tequila.

Which they did. Lots.

It led to the kind of fun that only techies can appreciate. At the rented condominium, James and Michael competed to figure out how to get the gas fireplace to light. James had triumphed over the 3Com

AB8000, but was unable to light the fire. Everyone else slept, but James worked on the problem all night long. In the morning, the fire roared. He was still awake. Now he was tackling an infrared keyboard he had bought that wouldn't work with his portable computer. After six hours, it was up and running, too. James finally slept for fourteen hours.

Before they left Monterey, Edward asked each AsiaInfo manager to write down the ten most important things for the company's future and James tallied the answers on an Excel spreadsheet. There was a consensus about the number one priority. The company needed to replenish its coffers.

AsiaInfo's first outside investor was a friend of Edward's from Texas named Carol Rafferty. She put up $500,000 of her own money in 1995. After the Monterey retreat, Rafferty, who advised companies about fund-raising, said that she would help Edward write a private placement memorandum. He thanked her, but said, "We have one."

She read the two-page document and said, "Well, it's a start."

Besides helping with the memorandum, Rafferty counseled Edward about meeting with investors. "It wasn't difficult to believe in Edward," she says, "but he had to sell the company and China. In particular, China. The country was committed to development, so the presentation had to make investors understand that AsiaInfo was uniquely positioned to help. Could they deliver?"

Edward traveled to San Francisco to meet the friend of a friend named Peter Joost, a respected investor who managed the money of the Bass billionaires. Edward bought a new suit for the meeting and was wearing it when Joost's driver picked him up at his hotel in a black Mercedes. Edward met with Joost and an associate named Henry Lee in an office at the Bank of America building. A waiter rolled in a white-clothed table with serving plates covered with silver domes and a candelabra. When the waiter removed the domes, there were scrawny sandwiches. An elegant pitcher was filled with ice water. That was it.

Joost and Lee listened for awhile, but Joost stopped the presentation to say that he never invested in pure Chinese plays. Later, Henry Lee recounted the meeting in the *Wall Street Journal*. "Edward was a total visionary," he said. "He was talking about rebuilding China. There was nothing practical about him."

Throughout early 1997, Edward met other potential investors. In April, at the first conference devoted to the Chinese Internet, he was introduced to Bo Feng, who said that he was an investment banker with Robertson Stephens. Edward admitted that he didn't know exactly what an investment banker did. Bo explained, "We finance growth." He told him about the SRS deal.

When Bo asked about AsiaInfo, he couldn't understand Edward's technical explanation. He was nonetheless impressed when Edward noted that his company was the largest customer of Sun Microsystems in China. They agreed to meet again at Edward's office in AsiaInfo's Zhongguancun HQ. When they did, they talked for hours. Bo asked numerous questions until he understood the company's technology. When Edward showed Bo a small "strategy room," Bo noticed a wall-sized map of China with colored pushpins stuck in populous cities and far-off provinces. Bo was impressed; it was like a general's map in a war movie. Bo told Edward that he thought he could help AsiaInfo find "value-added" investors—investors who could contribute more than money. "Your company fits my criteria," he explained. "I'm not interested in a company even if it has a great product, a great technology, without a vision. I'm not interested in a body without a soul. I am interested in companies that will live on and make a mark. The leaders must have more than an understanding of the market opportunities and a competitive advantage. They must be interested in creating an organization that produces for the long term."

They talked more at dinner and discovered how much they had in common. They had similar memories of the Cultural Revolution. They discussed their experiences in America. In addition, both of their lives changed because of Tiananmen Square. When they discussed June 4 and Edward heard about the black ribbons on the trees, he looked closely into Bo's eyes. Their friendship began in earnest that day.

In subsequent meetings with Edward and his colleagues, Bo discussed options for funding AsiaInfo. When they asked Bo how much AsiaInfo was worth and how much equity they would have to give to investors, Bo said that he needed more information. "I need your P/E and EBITDA," he said.

The founders looked at one another. They didn't know what he was talking about.

Bo explained, "P/E is your price to earnings ratio. EBITDA are earnings before interest, taxes, depreciation, and amortization." He showed the formulas on Edward's yellow legal pad, explaining why the numbers were essential.

Before Bo committed to AsiaInfo, he wanted Sandy Robertson to meet the AsiaInfo founders, so he arranged a meeting in San Francisco. With Edward, he flew to California, where they headed directly to the financial district and the impressive Robertson Stephens office in the Bank of America tower. The meeting was held in a conference room with Sandy Robertson and his partner and cofounder, Paul Stephens. James Ding was in Beijing, standing by on speakerphone.

To begin the meeting, Sandy asked Edward about the company's business model. Bo's instruction hadn't included anything so basic—he assumed Edward knew at least that much. Edward nervously yelled into the speakerphone, "What's our business model, James?"

From across the ocean, James's voice came weakly out of the tinny speaker box, asking, "What's a business model?"

Paul Stephens walked out of the meeting.

It could have been the end of the story, but Robertson was patient. "Your business model goes like this," he said. "A dollar comes in. You want to get a dollar or more out. How do you do that?"

Robertson also asked other straightforward questions. What are comparable companies? They didn't know. What's the potential market size? They didn't know. However, when the conversation went from the micro to the macro—that is, when Edward and James had the opportunity to explain their vision of the future of the Internet in China—Sandy understood what Bo saw in them. "After meeting them, I felt that it was a fairly low-risk deal. I was able to get out of them their sales figures, and their sales were good," he says. "They had a business plan even if they didn't know it. Regardless, I was engaged by their commitment. This business is about picking people. I sensed their promise."

Emboldened by Sandy's approval, Bo returned with Edward to China and helped him, James, and Yadong write a business plan section by section, dividing up the work. Bo advised James about the legally required component of the plan that explained the "risk factors," a feature of every business plan, but when Edward proofread it, he became indignant, running into a meeting with Bo waving the paper in the air.

"Why are you saying bad things about AsiaInfo! You are saying that we will fail!"

Bo calmed him down.

While Edward, James, and Yadong were happy to be in such respectable hands as Robertson Stephens, they also fretted that the company might try to take advantage of their inexperience. Their paranoia about the cost of the arrangement showed itself when, during a telephone call from Bo in California, Yadong snapped, "Hang up! You're spending *our* money."

When the business plan and supporting documentation were ready, Bo began his search for investors, relying on his contacts in the United States and Asia. On their initial road trip together to Hong Kong, Edward did a dismal job presenting the AsiaInfo story. Afterward, Bo said, "Don't read the business plan. Just talk. Be yourself. When you are, people will be convinced." Indeed, Edward got better, but not enough to win over any investors. One from Trans Bay, with impressive credentials, abruptly told Bo and Edward that he didn't think that AsiaInfo would fit into his portfolio; his company's most profitable business in Asia was a Taiwanese pig farm. "I know bacon fat and pork sides," he said. "I don't know what you're talking about." At a meeting at the Hong Kong office of GE Capital, Edward gave an impassioned speech, and at the finale was asked, "What's the Internet?" Edward wearily looked at Bo, but nonetheless embarked on a lengthy explanation, after which the man said, "Ah ha! Now I understand. You are a *plumber*."

Edward shrugged. "I suppose I am."

Edward wondered if this was a terrible mistake. Maybe he didn't want to take outsiders' money anyway. Maybe he should look to alternative sources of funding. What was he getting himself into? At his worst moment of panic, he told Yadong to fire Bo.

"Me?" Yadong asked.

"You."

"Why me?"

Edward responded, "It's your job. I'm not good at this sort of thing."

Yadong met Bo and informed him that the company had changed its mind, which devastated Bo. "I have spent four months helping you!"

he yelped. "You are the most unprofessional people I've ever seen! This is crazy! What are you thinking! Do you want to make this company a success or not?"

Yadong said that the decision was made, but it was canceled in a day; this time, Edward called himself. "This is terrifying for me," he admitted. "You have to forgive me." They met for beers and stayed up most of the night talking about how the Net would set the stage for a thousand new Chinese businesses. Before they retired at dawn, he agreed to go back out on the road.

Edward and Bo returned to Hong Kong for a second round and presented to respected investors such as Warburg Pincus and Newbridge. After one moving speech, Bo was certain that the investors in the room would clamber to write a check. However, when he looked closely, he saw that the top man was asleep behind his dark sunglasses. Still, the work paid off when in late 1997 an impressive list of investment companies, including ChinaVest, Warburg Pincus, and Fidelity, wrote checks for a total of $18 million. Sandy Robertson himself was in for $1.5 million. Bo earned more than $1.4 million in commission for Robertson Stephens.

The sizable fee irked the AsiaInfo founders. It seemed exorbitant, and Edward and Yadong decided that they wouldn't pay. When Edward asked a friend for advice, the man said, "Edward, you have to pay the bill!" but Edward responded, "It's just *Bo!* One point four million dollars for Bo? That's ridiculous."

When Bo got wind that Yadong had been assigned to negotiate a lower fee, he reported in to Robertson. "Sandy, I think they're going to try to dog us on the bill," he said. Sandy said that Bo should invite Edward and Yadong, who were in San Francisco to close the deal, to his home. "We'll celebrate," Sandy said.

The group arrived at Robertson's century-old Russian Hill mansion, where Sandy opened a bottle of wine from his cellar. They took their glasses out onto the deck, where they took in the view of San Francisco Bay. Afterward, as foghorns sounded, they walked down the steep hill past clanging cable cars. The bejeweled Golden Gate Bridge was lit up in the distance. At Bay Street, they ducked into Chez Michel, a French restaurant, where Robertson had reserved a corner table. Sandy proposed a toast to the deal while slipping Yadong the bill that

Bo had been carrying in his breast pocket. Yadong never mentioned the fee. Nor did Edward. Sandy never allowed the subject to come up and, when they were in the taxi on the way back to their hotel, Edward told Yadong, "I guess we have to pay."

Sandy and Bo stayed up late talking, laughing, and drinking wine. Sandy expressed some sympathy for the AsiaInfo bosses' feeble attempt to cut the fee—he had experienced similar stunts in the past—and then he turned serious. "I just want to say that you have done very well in a short amount of time," he said. "I think you have a knack for this business." In truth, Bo himself was stunned that he seemed to be pulling it off. He had no reason to believe that he could succeed in any business, let alone such high-stakes deals.

CHAPTER 6

TO WALK ON THE WATER IN BARE FEET

Before he could stand up on a surfboard, Bo bought three fine boards by Stewart, a renowned shaper; a pair of Rip Curl wet suits; booties; a leash; rounds of surf wax; a hood for icy water; and a wardrobe of surfer's T-shirts, board shorts, and sneakers. The hearts of the owners of certain boutiques in San Francisco, Shanghai, Hong Kong, Beijing, London—wherever—must palpitate when he is in their radius. He buys clothes, gadgets, jewelry, toys for our children, and presents for Heidi and his friends. Bo has impeccable but never conventional taste. There's a bit of punkishness in the mix. He wears Prada shoes like slippers so that the backs are crushed down and rolled-up Gucci pants for a game of soccer on the beach. Bo fantasizes about owning a Formula One team or, better still, driving his own Formula One car. When China has a Formula One team, it will be a meaningful benchmark for Bo. His nation will have caught up with the West. As Bo puts it, "Homeboys need to have one, too. Maybe it's a street man's dream, but why not?"

Back home in California, when, at dawn, we hear the motor of a Ducati on our quiet country street, we know it's Bo. He wakes up at 3 or 4 A.M. because of jet lag. That's when he puts on head-to-foot black leather and hops on the motorcycle, as red and polished as a hard-shelled candy apple. The bike flies north across the Golden Gate Bridge, cutting off the freeway at Mill Valley, where Bo rips up the side of Mount Tamalpais. At the summit, looking west—toward China!—on a sparkling clear day you can see the Farallon Islands. Bo winds down to Stinson Beach from there and turns onto Route 1, the coastal highway, through Dogtown, continuing north to and around the pristine Tomales Bay, past dairy farms. He turns back at Hog Island, the oyster farm, and heads through Point Reyes Station and then westerly along Papermill Creek. He's killed enough time; he knows we'll be up with the baby, so at six or so he arrives at our house with coffee and pastries. He teaches our young son soccer moves, plies our teenager with CDs, and fills us all in on the latest in China. By late morning, however, the dark underbelly of jet lag knocks him insensible and he passes out on the couch.

One weekend in late 1997, Bo arrives at dawn, but not on the motorcycle. He reluctantly parted with the Ducati because he's going to be a father. Heidi isn't thrilled about the new Porsche—a metallic blue 911—but it is questionably safer. She takes solace in the fact that Bo hasn't been able to drive the car much, because it broke down the day he bought it and twice since then. He has dragged along his brother-in-law, Jim Gislason. Jim, with a wicked sense of humor, straight-up hair, and what Bo calls "Falling Down" glasses (black and thick-rimmed like the ones in the Michael Douglas film), is one of Bo's closest friends, though he would never admit it at this moment. He is green from Bo's driving and from having been subjected to the Red House Painters CD four times in a row.

The three of us strap our surfboards, which have been stashed in my garage, onto the roof racks of my station wagon and we set out for the beach.

It's a frustrating day on the water for Bo. The waves are big and the current strong, and he can barely make it past the breakers. He's exhausted. Before long, Bo heads to the beach. After the surf session, the three of us sit in a coffeehouse, where Bo and Jim spend the first half hour arguing about French movies. Bo loves them and says that the existentialist filmmakers saved him. Jim says they're a bunch of wankers, trivial and pathetic. He doesn't mean it, but he loves to enrage Bo.

Bo has an announcement. He says that it has finally sunk in that surfing requires fanaticism, and it is increasingly difficult for him to surf since he is spending more and more time in China. Today is his last time surfing, he says with complete seriousness. Jim says, "Give me a break. If you stop surfing now, you have to calculate the cost of all your boards and your wet suits and the rest of your shit and you will never justify having bought them. I'd put the per-wave cost at about sixteen hundred dollars per wave."

Bo disregards him. "Soccer is better for me," he says. He has played soccer since he was a teenager. Bo is a demon on a field. He says that soccer is much more satisfying for him, particularly since it can be played on both continents.

No one could ever accuse Bo of being irresolute. He never surfs again. He does try snowboarding on a weekend that he, Heidi, and Tiger spend with friends in Colorado. Enthralled, he buys the flashiest gear made, but never has time to try it again, either.

Back in China, Bo sometimes plays soccer on the weekend, but there's not much time to relax other than occasional weekend games of mah-jongg or poker. They typically start at the end of business, say 10 or 11 P.M. He plays with Baojun and other friends in intense, smoky games that can last eight or more hours. When Monday morning arrives, his friends head home to their beds, but Bo goes directly to his first meeting of the week.

In May 1998, it's with Wang Zhidong at SRS. Raising money is an investment bankers' job, but Bo doesn't hit and run. Over the course of the lengthy initial SRS and AsiaInfo deals, he has become close friends with Edward Tian, Zhidong, and some of their colleagues. After writing and rewriting business plans, nose-to-nose meetings, nearly being fired, calamitous road shows, and endless flights throughout Asia and across the Pacific, Bo is a virtual member of AsiaInfo's and SRS's management teams and a semiofficial adviser as well as confidant to Edward and Zhidong.

We arrive at Zhidong's office in Zhongguancun on a smoggy and searing afternoon. It's a small and stark room. There is no air and the open window is useless. A fan drones. We sit on metal chairs facing Zhidong's desk. Another man, Wang Yan, an SRS manager, leans on a metal bookcase. Wang Yan, who is not related to Wang Zhidong, is wire thin and has a sharp razor haircut. While Bo is unchallenged as the industry's classiest dresser, Wang Yan is its flashiest. A typical ensemble includes a canary yellow shirt with a green silk tie under a mauve sport jacket. Wang Yan grew up in Zhongguancun. "If you are from this district, you have a warped vision of the world," he says. "If you are sent to the store to buy apples, you must be reminded to buy real apples." Not computers. He left China to attend the Sorbonne in Paris. After earning a law degree, he returned to Beijing and began working for SRS in 1995.

Zhidong and Yan reveal a shocking plan to Bo, who attempts to hide his alarm. Zhidong says that he is thinking of dramatically shifting SRS's business plan, even though some people at the company think he is making a disastrous mistake. However, he says, an idea is keeping him up night after night.

"All right," says Bo. "I'm sitting."

The presence of Wang Yan indicates that the idea has to do with the SRS website, which has become one of China's most popular. It was originally launched in 1995, one of the first in China. SRSnet.com was a comprehensive site designed to support RichWin. There was a bulletin board for users to discuss bugs and fixes. Under Wang Yan, the site became a sort of Chinese ZDNet or CNET, a repository of all things computer-related. However, Yan was surprised by the conversations on the bulletin boards. "We thought our users would discuss bugs and relate technical tips," he says. "Instead, they talked about football, gaming, and the news." The topics posted by SRSnet users ranged from current events to the least-visited forum, a discussion started by Bo himself. Its irresistible title: "Modern Youth and Existentialism."

At SRS, Wang Yan worked to adapt the site as its users changed from techies to what he came to describe as "China's Internet generation." "The Internet unites them," as Wang Yan described them. "It defines them. They are the first generation to have a global perspective."

At first, the number of users was inevitably small, since few people were online in China. Still, SRSnet got a large percentage of those who were connected—as much as 30 percent. The growth, of course, was related directly to the progress of AsiaInfo. At the end of 1996, as ChinaNet expanded, there were two hundred thousand registered users. There were a half million in 1997.

That's when Wang Yan was charged with expanding SRSnet.com into a "virtual Chinese society," adapting the western model of Yahoo. Along with Alta Vista's search engine, he put up a slick Chinese character engine designed by Zhidong's brother Zhigang. He also added chat rooms, more forums, news, and multimedia information about Chinese culture, everything from movies and music to traditional theater. By late 1997, Wang Yan increased the coverage to include computer and video games and social issues, and he again added more news. He tracked the site's traffic and saw that it was coming from foreign as well as Mainland Chinese. The online discussions at SRSnet were free and expressive, at least by Chinese standards, though the company avoided governmental interference by walking a fine line of self-censorship. "We told users of the forums, 'DO NOT PASTE EXPRESSIONS AGAINST CHINESE LAW OR THE GOVERNMENT,'" Yan Yanchou later explained. "If anyone pasted these contents, in order to keep the survival of our website, we deleted them."

The Internet had already begun to place the Chinese government in a quandary. The government understood that the technology was essential to its goals of modernization and globalization, but it simultaneously threatened its control. "Recognizing that an unregulated network would shift power from the state to citizens by providing an extensive forum for discussions and collaborations, Beijing has taken steps to prevent this commercial goldmine from becoming political quicksand," wrote Nina Hachigan, a senior fellow at the Pacific Council on International Policy, in *Foreign Affairs*. "But a victory over cyberspace cannot be decisive because the Internet cannot deliver its full commercial benefits under strict political control." It's why Beijing seemed schizophrenic—by turn encouraging and then discouraging investments in Internet technology and installing and then dismantling regulations designed to monitor users and forbid certain content and communication.

At the end of 1997, the government's State Council released the "Regulations on the Security and Management of Computer Information Networks and the Internet." Included: "No unit or individual may use the Internet to create, replicate, retrieve, or transmit [anything that could] incite or overthrow the government or the socialist system. . . ." Technology could not be used for "making falsehoods or distorting the truth, spreading rumors, [or] destroying the order of society." "Promoting feudal superstitions, sexually suggestive material, gambling, violence, or murder" was forbidden. Anyone in the Internet business had to "accept the security supervision, inspection, and guidance of the public security organization. This includes providing to the public security organization information, materials, and digital documents, and assisting the public security organization to discover and properly handle incidents involving law violations and criminal activities involving computer information networks."

Since they were enacted, the regulations were enforced and ignored with seemingly random frequency. At first, service providers meticulously required every new user to fill out registration forms. When that discouraged people from signing up, the companies themselves filled out the forms and submitted them to the government. Soon the providers stopped filling out the forms altogether.

Regulations or no, Net use grew rapidly. (And it would continue to grow: By the end of 1998, there were 1.8 million people online, according to the China Internet Network Information Center, and the number of "net worms"—*wangchong*—would more than triple in 1999 to 7 million. A survey conducted that year showed that four people were using each registered log on, which meant that there already may have been as many as 20 million occasional users. By mid-2000, there were at least 20 million registered users and 35 million e-mail accounts. Since a dozen or more people sometimes shared one e-mail account, the real number was far higher. In addition, occasional users could buy a prepaid Internet card and log on anonymously at an Internet café, using Hotmail or a similar Web-based system and a fake name.)

Throughout the growth of the Net, SRSnet remained the most popular portal in China, breaking records for a series of events of interest to Chinese from around the world. In 1997, when the Chinese soccer team was in the World Cup preliminaries, SRS did real-time broadcasting and was packed with more than 3 million hits a day in May and June, a record for any Chinese site. That year, SRS allocated more money to enlarge the website and spun off a separate site to support its software products. However, there was a roadblock to growth. In separate discussions, Yan Yanchou told me, "We have the best and the biggest Chinese site in the world, but our users are mostly in China. There is a whole world of Chinese users who aren't being served." Wang Zhidong asked, "What about the Taiwanese, North American, and the other Chinese communities?" and his brother Wang Zhigang said, "We see a dream of uniting all Chinese people, not only those in the Mainland. We want to make ours the biggest Chinese site in the world. How can we reach out and unite the world's Chinese?"

Sitting up on his hard chair at their 1998 meeting, Zhidong tells Bo, "Some dramatic thinking is called for. We are a Chinese company, but we need to reach out further." He reveals his plan. "I am going to change SRS so that it becomes an Internet company. It will no longer be a software company."

Bo gasps. "What?" Unable to control himself, he blurts, "This will destroy SRS! What about everything you have built?"

Zhidong sips water from a plastic cup before calmly explaining, "It is the future."

"But RichWin *is* SRS," counters Bo.

"For how long?" Zhidong answers. "The software business is stable, but growth is limited to the growth of the PC user base. That's not bad, but the prices for software shrink and the margins shrink even more."

"Explain the business model," Bo says. "Are you thinking about Yahoo?"

"Some things like Yahoo, but a full website specifically tailored for China," answers Zhidong. "Who else can do it? Yahoo China is coming here, but a Chinese company should be the portal for our people. We understand China. We can give people what they need."

Bo doesn't say anything, but he is thinking, with no small degree of anxiety, that Zhidong is contemplating taking the largest software company in China and risking everything on an unproven business. SRSnet is China's most popular site, but it generates no revenue. Not a penny. Software companies are tried-and-true businesses. Portals in the

United States have high valuations (at this time), but with the exception of Yahoo, no profits are in sight.

Zhidong looks up at Bo with his heavy lidded eyes. "You amaze me, Bo," he says. "You are Mister Internet. You are the one who says that the Internet is China's salvation. China will remain behind the West if we don't leap into the future."

"I don't disagree with any of that," Bo says. "But there are lots of other Internet companies. Look how much money Microsoft has burned trying to figure out the Net."

"If we don't do it someone else will do it," answers Zhidong. "If we don't make it our only priority, someone else will do it and they'll probably do it wrong." He says that he has put in motion what he is calling Project Surf, an evaluation of what it will take to transform SRS from a software to an Internet company. He also admits that there is some fierce opposition at the company, although Wang Yan and Yan Yanchou are on board. (On the other hand, when he heard the idea, one manager renamed it Project Stinky Fish.)

Bo is cautious, thoughtful. "I want to hear more," he finally says. "I don't have to tell you that it's a different environment here. In America, Yahoo has the support of a far more mature industry of partners who pay for access to the huge Yahoo customer base. In addition, there is a much more sophisticated advertising industry in the U.S. As far as I know, Yahoo's revenue comes from those partners and advertising. What's your vision of growth, and what are your projections of revenue?"

Zhidong, Wang Yan, and Bo continue talking as we walk down a hallway. We enter a low-ceilinged room with tables pushed together in a U, where Yanchou is working in front of a PC. There are a dozen workstations on long tables with their cables snaking along the floors and tacked onto the walls. Yanchou wears a rumpled short-sleeved plum shirt and long pants. Because of the heat, he has rolled up his pant legs above his knees. His face, with sloe eyes behind his metal-framed eyeglasses, wide cheekbones, and a furrowed brow, has a bemused smile. Yan, who is fifty-five years old, occasionally spends eighteen-hour days at game bars, playing computer games, but now he is working diligently on Zhidong's new project. He is soon echoing Zhidong's enthusiasm about the Internet. "We will unite Chinese speakers throughout the world," he says, "thereby creating an online community such as the world has never seen. It's the step that's required in our goal to bring China into the center of the global economy, because our people require the same information and same immediate ability to communicate as their peers in the rest of the world."

Bo thinks, This is terrible.

The group heads out to lunch, piling into two cars that are waiting outside. We drive through Zhongguancun and screech to a stop in front of the crowded, ornate Duck Restaurant, where we are led to a private back room painted lime green. When a cell phone rings, each businessman flips a phone out of his respective pocket, but it is the waiter who is getting a call.

We are served soup in lidded white bowls with blue rims and glasses of jasmine tea followed by the restaurant's namesake, Beijing duck, with sweet sauce and crispy skin, served with strips of scallions and puffy rice-flour crepes. After complaints about the weather, the group discusses the marketing and licensing of the latest version of RichWin for the Internet, which does for Net browsers what RichWin did for the operating system. It recognizes and displays correct Chinese characters and includes a cursor dictionary so that the user can click on an English word on a website and instantly get the Chinese translation. They discuss management issues, too; Zhidong asks Bo if he can help recruit for some key positions. "Good managers are impossible to find," Zhidong complains. "The shortage is worse than in the West." Bo agrees. "There are plenty of dreamers with ideas, but little sense of the ways to manage organizations as they grow from sales of a few million dollars to hundreds of millions and even billions of dollars." (It's why Chinese premier Zhu Rongji will show up in April 1999 at MIT after graduation to invite Chinese students to return home, stressing that China's greatest need is management expertise.) "Managers in China must have a broad skill set, plus adaptability," Zhidong says. "Everything, from corporate hierarchies to incentives to the motivation of workers, is evolving here. One needs an incredibly fluid management style that isn't held back by preconceived ideas about the way things have to be. From the West, one can fine-tune some ideas about teams and incentives. However, Chinese people have to be taught to understand them. From China, one needs to understand the local business community and the history if you want to get anything done."

Next Zhidong raises Project Surf. Wang Yan reports, "There are sixty thousand new registered users a month. The time has come for us to make a decision about our direction, however. We can either remain in the lead by offering our users a complete website that will become the focal point of their lives as they enter the Internet age, or we can let someone else do it."

Zhidong surveys the room. His captains are behind him, though Bo isn't yet on board. He says, "We're evaluating what it would take to make this our highest priority. It's a very big decision and I'm ready to go forward."

CHAPTER 7

CHINESE CAPITALISM WITH WESTERN CHARACTERISTICS

The open windows in the taxi blow in air as if from a hair dryer. The heat brings on a lethargy in Beijing, but not in Bo, who is rushing to his first meeting of the morning. It's with a guy who has come up with an "operating system overdrive"—basically, software that makes computers run faster. It's called Doors.

"Doors?" Bo repeats.

"Doors."

"Like Windows?" Bo has a deadpan expression, amusement in his peaked eyebrows.

"Windows?" asks the man. "What do you mean?"

Bo inspects the man. He says, "The image I get from 'doors' is a closed door. Is that what you want to project?"

The man, who is writing down every word Bo says—"*Is that what you want to project?*"—looks up and answers, "Doors open, though. My program opens doors. Don't you like it?"

We visit Edward at AsiaInfo. The company has soared since Bo's Robertson Stephens deal. ChinaNet, built largely by AsiaInfo, has nodes in 230 cities in thirty-two provinces and administrative regions. Service may be expensive, but prices are dropping. ChinaNet remains the primary network for most Internet traffic in China, and it is among the largest commercial networks in the world. Virtually every service provider operating in China routes some portion of its traffic over ChinaNet. AsiaInfo built it. The Internet has reached China.

Edward has learned an enormous amount in the past year. After the Robertson Stephens deal, Edward chaired his first official board meeting with the newly seated representatives of his investors. The day after the meeting, one of the new board members messengered over a copy of *Robert's Rules of Order*. He now runs meetings with professionalism, and he has honed his innate ability to inspire and elevate those in almost any room. After his disastrous "infomation" scheme, he returned AsiaInfo to what it did best—wiring the nation. In addition, AsiaInfo is increasingly being asked to get back into the Intranet (in addition to the Internet) business. As WTO negotiations continue, it is becoming increasingly clear that Chinese companies need to automate if they are going to survive in an environment of real, rather than managed, competition. Diverting resources to systems integration now makes sense; Edward was just too early by a year. In fact, corporate China's push to reengineer and computerize is beginning to fuel what will likely be a systems integration boom unlike any the world has seen. AsiaInfo is winning most of the major contracts related to the telecommunications industry. Edward and James's growing reputations make an enormous difference. So do their carefully managed relationships with the government agencies regulating the Internet and telecommunications industries. Contracts in China are only now beginning to be subject to open bidding, which is a roadblock faced by many of the much larger competitors from abroad. As Bo says, "AsiaInfo's technology solution is unsurpassed and its good relationship with the government ministries means that it can make things happen here."

AsiaInfo is poised to profit from the myriad ways the Net, and perhaps other communication systems, are growing. The company continues to work on China Telecom's ChinaNet, does major systems integration deals with private as well as government companies, successfully sells a business software package it created called Userfriend, and creates and maintains corporate Intranets. Edward estimates revenues of $1 billion for 2000, up from $40 million now in 1998. A major deal was made with China Unicom, an upstart telecom that focuses on wireless telephone and paging; AsiaInfo has been hired to build the company's internal systems. *Forbes* notes, "The company now has a virtual monopoly in building Internet infrastructure for Chinese telecom companies."

Zhongguancun is packed when we snake through the streets under bright red, yellow, and green streamers. When we reach AsiaInfo, we find Edward in a meeting behind closed doors. When it's over, he joins us for a walk along Haidan Lu. Edward tells us that Jean is pregnant. He says, "I want to make time to get to know the new baby." It's unlikely, though. He shrugs off the thought by saying, "I'm inept when I'm home anyway." He tells us about his last visit to Dallas. Trying to be useful around the house, he was pulling weeds in the backyard. "I got frustrated and decided it would be easier to burn out the dead grass." The image is reminiscent of Edward in the desert conducting research on the impact of burns on the grasslands. He lit a patch of weedy grass with a match. Within seconds, a fire blazed, engulfing most of the backyard, approaching the wood fence. Edward grabbed one of Stephanie's buckets and ran to a spigot, filled it with water, and dumped it on the flames. He recalls, "When Jean ran outside, she looked at me with the toy bucket, running back to the house for more water, like she couldn't believe what I was doing." She called 911.

Edward laughs, but he knows it's not all that funny of a story.

Bo says, "Stick to running companies."

We continue our walk. The sidewalk is busy with foot traffic and bicycles. It's noisy, so Bo has to speak loudly. It's the first time he reveals to either Edward or me that he is unsure about his job.

"What is it exactly?" asks Edward.

"I'm becoming frustrated as a banker," he says. "If it wasn't for Sandy, I would quit."

"Isn't this your dream?"

"Investment banking is a useful job," he says, "but the time isn't right for that career in China. A lot of the companies I'm looking at aren't ready for an investment bank to bring in ten million dollars. They need seed money—a half a million in order to figure out who they are. In addition, bankers make the deal and then move on to the next one. The investors themselves are expected to stick around for what is the true challenge: to build the companies. The investors become board members and advisers, whereas I am off chasing the next deal. It's not that capital isn't key. But Chinese companies need more."

Bo continues: "China needs hundreds of AsiaInfos. For the start-ups I'm visiting, money is the least important requirement. They need help. A lot of companies make you guys, back when there were ten of you, seem like Cisco or Apple. They need *a lot of help*."

Edward asks, "What are you going to do?"

Bo says that ChinaVest has offered him a job. Edward knows ChinaVest well; the VC company is one of the investors Bo found for the AsiaInfo deal. It's a respected firm, the first VC company to invest in China, though most of its investments are in old-economy companies. ChinaVest has funded manufacturing companies, restaurants, and pharmaceutical companies; its portfolio is diverse, including a joint venture in China and Taiwan with Domino's Pizza, T.G.I. Friday's, and Tait Asia, which distributes Heineken, Disney videos, and Viagra. Bo says, "I'd be in charge of helping the firm to move into information technology. The pay is good, and I'll have the freedom to pursue interesting deals." He adds, "But I won't do anything until I discuss it with Sandy."

The next morning, after an extravagant three-and-a-half-hour nap, we board a train to Shanghai. Our "soft" seat costs $6. Train travel is ideal for Bo because he can spend the entire time on the phone. (There are no telephones on China's domestic airlines.)

Hot air pours into the open train-car window. We shoot like an arrow through parched fields, past ramshackle apartment buildings, a crumbling car-parts factory, and a vacant schoolyard gold with blooming *jinju* (chrysanthemums). Bo doesn't notice. He is preoccupied. He finally says, "Balance is everything, but balance isn't my . . ." He elides the thought. In a moment, he continues. "The problem is that balance is not my nature." It's an understatement. Bo becomes thoroughly immersed in whatever new idea or pastime or person captures his attention. It's partly why he has succeeded so quickly, but it's difficult to carry on a personal life if one is obsessed.

The subject makes Bo somewhat anxious, since it raises an unresolved, probably unresolvable issue. He's uncertain that he can pull off the life he has constructed for himself. It's not whether or not he can do the work. Neither is it the weight of his growing responsibility. It's not even the fragility of China or whether he can make the time required for his familial relationships. It's a more subtle conflict. Bo is twisted up about who he is somewhere between China and the United States. It is almost innate in the sharp clash of the cultures that rule him. Bo's blessing and his curse is the fact that he is Chinese with an American life at least half the time. Dividing his time between both cultures and married to a woman from California, he is always torn—missing one, frustrated by the other, compelled to act Chinese, American, Chinese again. It's obvious that he adores his American friends, but he is beyond passionate about China. Yet even those feelings are conflicted, swinging between pride about his nation's breathtakingly beautiful culture and frustration at its aggravating backwardness. Shen Baojun says, "It takes a great deal of effort and the transitions are not always seamless." Indeed, the more time I spend with him in China, the more I understand how difficult it is for Bo. It is a sign of his intense discipline that he pulls it off. However, the subjugation that's required, and the internal stress, take a toll. "There is always one cuff link on the other side of the ocean," Bo says. "The matching sock is in China when I am in the U.S., in the U.S. when I am in China. To solve that, I have bought many pairs of the same cuff links and piles of the same socks. But I can't have two Heidis. Sometimes, without her, I feel as if I could lose myself. And sometimes, when I'm away from China, I feel like a fish trying to live on the land."

The train passes apartment houses with laundry on lines and boys playing soccer in a field. There are small gardens and acres-long fields of corn. We rush by a metalworks with cracks in its concrete walls, a slow-moving oxcart, and a long, straight line of sunflowers. When we slow down, nearing Shanghai station, there is a thick throng rushing toward the train, pushing in. The doors open, and Bo and I make our way through them and leap into a cab, which brings us to Bo's office for the day: a humble bar with walls made out of corrugated iron, with silk curtains, Chinese opera from an old radio, and a string of blue, yellow, and red Christmas lights hanging from ceiling rafters. Behind a bar stands the rotund and scrappy owner, who pours beer into dirty glasses and serves plates of mint-flavored beans, spicy dried peas, and Pringles.

More entrepreneurs, one after the next, take the stool opposite Bo. The presentations consist mainly of dry, monotone laundry lists of technical details and numerous and tedious PowerPoint slide shows. One man wants to set up Internet terminals in streetside kiosks. One wants to start an equivalent of *Wired* in China. There's a plan for an online matchmaking service—"matchmaking workers, jobs, couples." A nineteen-year-old prodigy who attends Beijing University is developing marketing software that personalizes Chinese e-commerce sites, replicating a service offered on Amazon.com in the West. (The software recommends products based on past purchases plus one's Internet surfing.) Bo is ever forbearing in the face of doomed presentations, but there is an occasional gem. You never know.

"Who is your competition?" he asks the designer of a new type of translation software.

The meek man, who is only twenty-two, says, "I don't know."

Bo asks, "Do you think you need to find out?"

The man scratches his head and pushes up his bifocals. "I suppose so. How?"

Bo looks at me and shakes his head before embarking on an explanation.

Bo has boundless and infectious drive. His wiring doesn't allow him to sit still for long. Though he may be in a chair, he's moving, taking notes, dialing. Still the entrepreneurs come until the bar closes at midnight. And there's a plane to the States to catch in the morning.

If there was any doubt beforehand, last night's meetings confirm for Bo that he is in the wrong job. The reason: Even if any of the entrepreneurs he met had promise, they are years away from the established positions of AsiaInfo or SRS when Bo made his deals for those companies. Robertson Stephens wouldn't touch them. He thinks about it over the course of the flight to San Francisco. There's no longer any doubt in his mind—as long as Sandy agrees.

Bo taxis directly to Sandy's office from the airport. They greet one another warmly. When Bo tells Sandy about the ChinaVest offer, Sandy responds with other news. Robertson Stephens is being acquired by the Bank of America. Sandy explains that BA has its own plans for China. "They're hesitant about much investing in the Far East," he says. "But they want you to stay with the company. They have a position for you in Hong Kong."

"What's your advice, Sandy?" Bo asks.

"It may well be the time to try something new," says Robertson. "Tell me more about the ChinaVest offer."

When he hears the details, Sandy encourages Bo to go to ChinaVest. He says, however, "One way or another, I know we'll work together again."

Like a boomerang, Bo is flying back to China, where he meets Jenny Hsui Theleen, ChinaVest's cofounder. ChinaVest, with $450 million of its funds invested in China, is located in the office building associated with the Swissôtel in Beijing. They meet in the bar. Theleen agrees with Bo about the importance of VCs in China at this particular moment in the nation's history. "VC money is more valuable in China than ever," she says. "Since the Asia crisis of 1997 and '98, the public markets are so bad that venture capital has become the only source of funding for growing companies." They negotiate the details and it's done. Theleen seems excited to have landed Bo. "He has closed two of the biggest technology deals in China," she notes.

In January 1998, Bo officially leaves Robertson Stephens and begins his job at ChinaVest. He has the backing of a large and well-respected firm with enormous experience in China. The job is straightforward: find solid investments of the caliber of AsiaInfo and SRS. In so doing, broaden ChinaVest's portfolio in high tech. He is ecstatic. This is what he is waiting for. He sets off to more meetings—dozens of meetings with entrepreneurs each week—searching for the next Edward Tian, the next Wang Zhidong.

In summer, another transpacific flight brings Bo back to San Francisco in time for the birth of his son on July 6, 1998. The baby is accommodating. Bo has two days to acclimate before they rush to the hospital. It's the Year of the Tiger, so Heidi and Bo name their son Xiaohu, "Little Old Tiger." Bo, who early in the pregnancy attended Lamaze classes with Heidi, dons a green hospital gown and experiences what he describes as the "penultimate experience of my life." Afterward, he seems so ecstatic that we wonder if he might faint. Heidi and Bo bring Tiger home and adapt easily. Bo's mother, Lihui, arrives in San Francisco two weeks later. Heidi, who resigned from her job as assistant director of the UC-Berkeley Public Service Center, has her hands full with Tiger and the renovation of their new apartment. When it's time for Bo to go back to work, he feels torn. "All I want to do is play with Tiger," he says. However, China is calling.

CHAPTER ★ 8

THE BUSINESS OF A REVOLUTION

A chill shoots down my spine when I read an article about China one morning in late 1998. Wu Jichuan, China's information minister (he heads the Ministry of Information Industries, or MII), has released a sweeping statement about the government's latest plans for the Internet. Minister Wu seems to hover over the Chinese Internet entrepreneurs like a specter, the personification of the government's potential to squash them. Wu is discussed both with a begrudging respect and a pale enmity because of his enormous power—enough to make or break a company, even cripple an industry. It's why he is notorious and feared. Internet consultant Duncan Clark notes that Wu, minister since 1993, is from Hunan and as such is "a spicy character" who is unpredictable and potentially dangerous. However, Clark says, "falling victim to the Patty Hearst syndrome, the Internet community, which has been hostage to him for so long, has fallen in love with him. We live and breathe for his blessing."

In his latest statement, Minister Wu says that the government will enforce the standing requirement of Internet providers to register their users and will hold them responsible if their users break the rules about the allowed types of online conversations. Wu says that the government is cracking down on the increasing flow of Western news into China and the ministry is looking into the "growing problem" of foreign investment in Chinese Internet companies.

I frantically call Bo, who seems amused by my concern. How can he be nonchalant about Wu's statement, which seems to be a direct attack on his work? In fact, the "growing problem" of foreign investment in Chinese Internet companies could be read as the "growing problem" of Bo's success.

"It's normal," Bo says. "China is sorting things out."

"But the government is blocking everything you are working to build!"

"Remember I explained how old Chinese coins are round on the outside, but square on the inside? So it is with everything in China. Things are not what they seem on the outside. The truth is hidden."

What an exasperating answer! "Which means?"

"Which means don't worry," says Bo. "There is a different story underneath the words. In China, many voices speak at once. There is a hidden reality that exists and moves along at its own pace. We can't worry about all the voices. We just do our work."

I'm far from reassured. Is Bo really so certain that there isn't anything to worry about, or is his calm an example of the circle on the outside of a coin? I conclude that after a lifetime in China, Bo knows what I will come to learn. As he puts it, "Every day there are new fires. If you spend all your time trying to put them out, you'll never get anything done."

The regulations on the books designed to control Internet content, communication, and investment are baffling to outsiders. There is a range of opinions about them. A businessman in Beijing tells me that regulations in China are "like bamboo—stiff but bendable." Many of the government's pronouncements, which make for breathless headlines in the West, are viewed in China as posturing by government agencies, testing the waters, or perhaps "land grabs" for control. Indeed, Minister Wu at the powerful Ministry of Information Industries is only one of the agency heads attempting to exert control over the Internet. The Ministry of Culture has barred websites with foreign investors from selling audio or video products, the Ministry of Education requires preauthorization of online schools and education portals, and the State Press and Publishing Administration has demanded that book retailers obtain licenses to operate online. Most of this is ignored. The Ministry of State Security, however, is tougher to discount. It is the agency behind the arrest of Lin Hai. Nor is this to say that the other government agencies, particularly Minister Wu's, aren't forces to be reckoned with. However, in spite of their power, Clark says that Bo is correct to remain focused on his work and not to worry about the pronouncements—for now. "Many announcements just mean that there is a conversation going on," Clark says. "Minister Wu, for example, will make a statement and the contrary arguments will flood in. The government will weigh them and decide on a course of action." The problem is the unpredictability and precariousness. Laws are enforced, if sporadically, and some regulations have teeth.

In 1997, prior to the arrival of Chinese president Jiang Zemin to Washington—the first state visit by a Chinese leader in twelve years—President Clinton made an address that was broadcast over the Voice of America. "The Internet already has 150,000 accounts in China, with more than a million expected to be online by the year 2000," he said. "The more ideas and information spread, the more people will expect to think for themselves, express their own opinions, and participate. And the more than happens, the harder it will be for their government to stand in their way. . . ."

It's what the Internet community counts on. Still, it takes awhile for me to become convinced—in spite of the idealistic visions of people like Bo and Edward. I can see how the Internet can bring change by democratizing information and communication, but only if it is allowed to grow unfettered and uncensored, which seems unlikely. Yes, it can help the ailing Chinese economy, providing new-economy jobs in a country that has relied on inefficient and unsatisfying make-work in state-owned enterprises, but I am not sure that new technologies can dent the enormous numbers of unemployed and underemployed. I also wonder if information and communication and even a degree of prosperity are enough to fix China. Michael Chase, who with his colleague James Mulvenon of the Rand Corporation is researching the Internet in China, told the *New York Times*'s Thomas Friedman, "If you just have information, but can't use it for organization or action to promote change, its political impact will be limited, and for now the Chinese government is still very good at preventing people from turning information into political action." However, whereas some conversations I have in China are stilted with distrust, weariness, and confusion, most are indistinguishable from conversations I might have with people in other parts of the world. Many, many people I meet are as boldly opinionated as any American. It's not what we Westerners have been taught to expect. In his *Times* piece, Thomas Friedman expresses the countervailing argument, too. "Yes, it's true that the Chinese government has tried to block [Net] access, but it's not working," he writes. "Deep down, the leadership here knows that you can't have the knowledge that China needs from the Internet without letting all sorts of other information into the country, and without empowering more and more Chinese to communicate horizontally and create political communities. In the long run this will only give more tools to the forces here pushing for political pluralism."

The potential number of Internet users in China is staggering. For the near future, it would be a mistake to count most of China's billion-plus people, since many of them are impoverished and isolated. Nevertheless, President Clinton predicted a million Chinese online by 2000, but the number turns out to be more than 20 million. Depending on who is doing the estimating, there may be between 33 and 60 million users by 2003, and 100 to 140 million registered users by 2005. In 2006, if not sooner, China will likely have the world's largest Internet user base.

The population will come online via computers—their *dian nao,* or "electronic brains"—plus set-top boxes (TVs that double as Internet terminals), personal digital assistants (Palm Pilots, etc.), and cell phones. "For many Chinese who work on the move, there may never be a need to get a fixed line," says Bo. "They will go from no access to complete, portable wireless access." An analyst recently said that mobile phones in China are as common as "bamboo shoots after a spring rain." By 2005, the country will have an estimated 200 million cell phones, most of them Net ready. Similarly, Edward Tian points out that most Chinese will skip past the early years of frustratingly slow Internet access over modems and leap directly into broadband. "Broadband in the U.S. is growing slowly because there is the cost and resistance of older technology," he says. "In China, most people's first connection will be over fast pipes." There are estimates of 100 million broadband connections in China by 2004, maybe sooner.

What about the rest of China? The potential for a Western-style digital divide between Internet haves and have-nots is enormous in China, but some Internet entrepreneurs believe that their country will solve the problem faster than the West. "Because cities that never had phone service are being wired for broadband, before too long I think we will basically wire the nation," Edward says. "We will wire China. I'm thinking three hundred million Internet users in five years. In ten years, the technology will be accessible to most of the nation—at least as many people as the number with televisions, which is eight hundred million."

As significant as the numbers of users is what the Chinese are doing on the Net: mostly, just what the rest of the world is doing. Pornography is forbidden, though it should surprise no one that Net surfers can get all they want. The most popular websites, portals like SRSnet and Sohu ("Search Fox"), help Net surfers conduct business, research everything from academic papers to recipes, meet one another, and, increasingly, buy.

Buy? Since only a couple hundred thousand Chinese have credit cards in 1999, a lot of the early Net shopping is being done with debit cards. The system is relatively crude. For one thing, many debit cards can only be used in the city in which they are issued. Until a better system takes over (IBM and Bank of China are working on one), many people pay for online merchandise with money orders purchased at a post office. Some websites offer COD delivery, and others require customers to make a payment by money order or a direct deposit into the websites' bank accounts. Customers are notified by phone or e-mail when payment is received.

Transporting goods bought and sold online is another obstacle. The mostly rural nation has no equivalent of UPS or FedEx, though both companies are ramping up their China operations. In the meantime, China Post, the state-controlled postal system, has a mixed reputation for reliably delivering packages, but it is spinning off a FedEx-like start-up. Biding their time until reliable delivery systems are in place, some of the first Web stores have solved the problem with simple technology: bicycle messengers. The website 8848.net.cn (named after the height of the Chinese face of Mount Everest—8,848 meters), which sells electronic equipment, software, books, CDs, and DVDs, uses UPS's rapidly expanding services. Users of Eachnet, a Chinese online auction house with hopes of becoming China's eBay, sometimes meet on street corners to consummate the deals they make online.

E-commerce—"the future of business in China," according to President Jiang Zemin—may be in its infancy in China, but the growth projections are huge. It only generated $8 million in revenue in 1998, but the number should hit $3.8 billion by 2003, according to International Data Corporation, an American market research company. Peter Williams, the Asia-Pacific representative of Deloitte Touche Tohmatsu, the professional services firm, told the *International Herald Tribune,* "The potential for Asia, and in particular China, to overtake the West in terms of e-commerce uptake is not a matter of if, but when." E-commerce in the rest of the world may have stalled, at least when compared to the enormous growth that was expected in the early years of the Internet boom, but it could be a bonanza in China. The reason: If a new, efficient system is established, it will replace a Byzantine system that discourages any commerce whatsoever. At many stores for many types of products, shoppers work with one clerk to select merchandise. When they decide to buy, the clerk fills out a form in triplicate. The customers must then move to a different station to pay and still a different one to pick up their purchases. Seamless online ordering would streamline shopping just as spendable income in China is growing sharply. Some industries that have been almost nonexistent or inefficient and disorganized, from banking to travel, have an opportunity to invent themselves on the Net.

Commerce is one thing—after all, buying and selling is a patriotic activity in China these days. But what about the fundamental changes to Chinese society that Bo and his friends envision? In the 1999 edition of the classic *The Search for Modern China,* Yale University China scholar Jonathan Spence writes, "China's emerging modernity reflected itself in additional material, rather than political, freedoms." The point: In spite of the grand speeches by Bo and the other Internet entrepreneurs, can the Net really make much of a difference in the life of the people of China?

In some ways the Internet is an antilogy in China. It represents freedom of expression and an open flow of information, whereas China has a long tradition of repressed speech and controlled information. Could the Chinese Internet be inherently different—encouraging buying and selling and government-approved information, but lacking true free speech, assembly, and exchange of information? It's possible to imagine a scenario in which the Net, rather than promoting freedom and openness, is used for propaganda and surveillance, a tool of a sort of technonationalism. One China watcher says, "A government that wants to watch its people must love cookies." (Cookies are the storehouses of information about a computer's user.) Privacy on the Net is an issue worldwide, but in China it raises particularly eerie possibilities. Could the government use the Net on a large scale to spy on individuals, read their e-mail, and track their Web surfing? James Mulvenon, who is an associate political scientist and the deputy director of the Rand's Center for Asia-Pacific Policy, told me, "Sure, the optimistic scenario says that the Internet's alternate channels of information begin to pluralize China. But . . . don't forget that the state can also use this technology for surveillance and to otherwise increase, not decrease, its control of its citizens."

I can also imagine another scenario, less nefarious but still troubling, in which the Internet solidifies, rather than undermines, the ruling Chinese Communist Party. If the government uses technology to help rescue the economy and the Chinese people have a better standard of living and somewhat more freedom, the citizenry may be less interested in any other changes, let alone a dramatic change such as anything resembling democracy. There are other possibilities, of course. Nina Hachigan, writing in *Foreign Affairs,* postulates that "the power shifts wrought by the Internet will surface only during an economic or political crisis in a future China where the Internet is far more pervasive. At that time, the Internet will fuel discontent and could be the linchpin to a successful challenge to party rule." To me, however, the more likely scenario is that the technology will continue to create a more empowered, more free-thinking, more critical population *and* a more responsive government. It will lead to an evolutionary, not an overnight, change. Of course, China is too big and the future too uncertain to make predictions with any degree of confidence. Even the present is ambiguous. However, the early indicators support the theory that the Internet is incendiary. One of the government's most effective tools has always been the control of information. Ann Beeson, an attorney who specializes in the Internet at the ACLU, says, "Every totalitarian regime everywhere attempts to control information and speech. By allowing in the Internet, [the Chinese government is] ceding control."

In spite of bans and blocks on any news other than the version preapproved by the government, there is a wide diversity of news, including foreign news, much of it translated into Chinese and accessible throughout China. Because there are so many conflicting reports about exactly what's available, I try Web surfing in Chinese Internet cafés, schools, office buildings, and private residences. With rare exceptions, I am able to access many of the sites that are officially banned. Self-censorship by the websites is sporadic, so Chinese sites may offer the *Times* or CNN one day and censor them the next. While links to foreign news media are illegal, state-controlled news is boring, so the sites often put up foreign news anyway. As Duncan Clark says, "The government's hand is on the lever, but the lever may not be connected to anything." One trip I am unable to access the *Times* of London, the *New York Times*, the *Washington Post*, CNN, *Playboy*, or dissident Harry Wu's Laogai Research Foundation, Laogai.org, about China's forced labor camp system. However, I reach Amnesty International's sites and a site devoted to Chinese dissidents in China and abroad (including speeches and writings by Wu and others). In addition, I easily surf to such Western news sources as *Time* (in spite of links to CNN), *Newsweek, Salon,* ABC News, MSNBC, and every U.S. government site I try. Neither do I have trouble reaching Taiwanese sites critical of the Chinese Communist Party nor Web pages devoted to Falun Gong, the Free Tibet movement, and the plight of Chinese political prisoners. Another time I try and get through to the *Times* and other Western media but not CNN and the BBC. The *New York Times* is blocked when I try to reach it in 2000, but it is accessible in the fall of 2001. The restriction was eased after the newspaper's top editors brought up the censorship of its site during an interview with President Jiang Zemin. In addition, I am easily able to surf from Yahoo China to U.S. Yahoo, where there is extensive coverage of the Tiananmen Square massacre—pictures, even news footage. Whereas my hotel has only CNN's business network on its cable system, I am able to read the complete day's news on sites based in Japan and other parts of Asia, as well as Yahoo in the United States. Overall, I find that there is less diversity of opinion and reportage on Chinese-language sites than sites in English, which means that Chinese who speak English have access to far more information. However, even the Chinese-language sites push every boundary.

Besides standard news, the Net is teeming with ideas and exchanges in a land where ideas and exchanges were, until recently, disallowed. It's why most people I meet in China aren't bemoaning the government's controls. Instead, they are astounded by what has made it online. E-mail is almost always uncensored, though there have been some troubling cases of the interception of e-mail by the security forces. Lin Hai's is the most iniquitous, since it led to his arrest. Chat rooms and bulletin boards are fiery. Filters sporadically weed out prohibited conversations, but for the most part they are ineffective. "Net mamas," whose job it is to police conversations, sporadically censor the conversation in chat rooms, but spontaneous and controversial dialogues in both Chinese and English are routine. Several translations of my own magazine articles about China have apparently made the rounds of the Mainland, because I have received hundreds of e-mails from people in major cities and tiny villages. In many cases the English was stilted, but the comments were free, curious, and opinionated. One e-mail referred me to a live discussion on a Chinese website where students from dozens of universities were discussing the Tiananmen Square massacre and debating the merits of democracy. "To go from no exchange to such a vibrant exchange is remarkable," Yan Yanchou told me. "Americans who are used to such an exchange cannot fathom the importance of this movement in China. It is a renaissance." While a few people have been arrested for posting banned news and opinions on their home pages, the enormous variety of free speech is impossible to monitor.

There's more. Whereas twenty years ago phone numbers in China were state secrets, the government always provided jobs and housing, and it was nearly impossible to relocate, freely travel, or change careers, the Net is a bazaar of opportunity and possibilities. On Netease.com, Chinese are selling mobile phones (for as little as $25). On job sites, Legend Computer, Newbaby, Motorola, and Friendly Crispy Chicken offer jobs. (Some require degrees and experience, some require English, but some say, "No experience necessary." One asks for "a bright smile.") Apartments are listed throughout China for as little as the equivalent of $20 to as much as $12,000 a month. A bulletin board describes the process for securing a visa to study in the United States or Europe.

That the government has waffled so dramatically on its Internet policies over the years makes it a dangerous, chaotic force. Will the flow of information increase, or will the controls tighten? Likely both. Which side will win? It's unknown, but I'm betting on the entrepreneurs. The reason is the technology itself. As a Beijing-based hacker once told me, "The Net is ultimately uncontrollable." The inconsistent messages about the Internet from the Chinese in part reflect the fact that Beijing is no less factionalized than Washington and that the government is sharply divided between those who fear and would control the Net and those who embrace it. China's security forces view the Net as a major threat, but the office of the premier is a strong supporter. In addition, different factions in Beijing have attempted to assert their control and influence at different moments. Regressive steps are countered by the progressive factions. Orville Schell says, "It's like a horse race between encouraging information technologies and resisting them by people who want to control information. No one knows which horse will win."

In the meantime, there's more cat and mouse. In early 1999, the Ministry of State Security orders the shutdown of one of the nation's most popular Internet forums when the open discussions center on the continuing crackdown on the country's nascent pro-democracy movement. In a week it also orders twenty-four-hour-a-day surveillance of online chat rooms; police are directed to "immediately contact the manager of the chat rooms whenever counterrevolutionary discussions are discovered. The customer's Internet service is to be terminated and the subject will be investigated." There are occasional closures of Internet cafés, which are popular from Shanghai to Dali, an outpost near Tibet. More restrictions on foreign news are announced a week later.

The intensifying climate is followed by baffling reversals. Few regulations seem to be enforced, at least for long periods of time. Closed Internet cafés are quickly reopened. The Net mamas that replaced government surveillance of chat rooms are largely ineffective; their presence seems to be more symbolic than anything. Banned Internet forums are reinstated within weeks or months, and there are few reports of surveillance of e-mail. Though the legal requirement to register Net accounts is now supposed to be enforced, China Telecom itself—the nation's biggest service provider and the government's own agency—continually ignores the regulation. When the government blocks or removes personal websites, new ones are up within hours.

The battle rages. One arm of the government attempts to control the technology, but some ministries and government branches are feverishly working with domestic and foreign companies to spread the technology to China's every corner. The state has unveiled a series of "Golden Projects," online systems for everything from tax collection to higher education. In 1998, the Government Online Initiative called for 80 percent of government agencies to be on the Web by 2000. In 1999, Minister Wu's MII forces China Telecom, the nation's main Internet provider, to lower prices of Net time so that more people can have access. Thus far there have been no attempts to limit the number of Internet service providers, though state-owned China Telecom owns most of the country's phone lines and dominates the early ISP business. In spite of that, there are a growing number of independent ISPs, and a rapidly growing number of Chinese homes and workplaces are connecting as the prices of computers, modems, and Internet services continue to fall. The number of Chinese online is doubling every six months. Schools are being wired, too. More than 20 percent of all Chinese in cities and a higher number in the countryside were illiterate in 1989, but the nation is dedicated to vastly improving the country's seven hundred thousand primary and middle schools. The Ministry of Education has announced that computer software and "cybertechnology" is a high priority in its goal of solving the education crisis. Xuefang Jin, deputy director of the ministry's Basic Education Department, says, "Computer-aided teaching will be necessary to reach the state's goal of educating one hundred percent of its children." A U.S.-based start-up company called ESS Technology signed an agreement with the

Ministry of Education to work on what has been described as "an education system for school and home for the twenty-first century based on multimedia and Internet technology."

The future of the relationship between the government and the Internet? The likely scenario is that some representatives of some government ministries and agencies will continue to attempt to control the Net with new regulations about content and investment. However, there's no going back now that the Internet is a national priority. I agree with James Mulvenon when he cautions about overoptimism. "The young urban wired in China think the world has changed, but when they get too giddy is just when they could get squashed," he says. "However, we'll see retrenchment, but fundamentally China is being altered by the information revolution." By building a state-of-the-art Internet for business, Beijing has set an irreversible course, ultimately unstoppable even if the Net begins to undermine the Communist Party's rigid control.

CHAPTER 9

CLICK DYNASTY

In the fall of 1998, Wang Zhidong arrives from Beijing at San Francisco International Airport, where he is greeted by Bo. He has a fuzzy bear for Tiger.

Bo drives the Porsche past broad fields of yellow mustard flowers in the Napa Valley and turns up a quiet driveway lined with old olive trees. Zhidong is telling him that he is ready to go forward with the transformation of SRS. Bo knew it was coming, but he can't control a sigh.

A bottle of '85 Martha's Vineyard works wonders to help Bo come around to Zhidong's way of thinking. They sit on the deck of the California French restaurant Auberge du Soleil and look out on the verdant vineyards. The air is dry and musty with the smell of grapes and fermentation. "If you are very cautious, it might work," Bo says. "Let's talk details. Software has a cost of production and a price in the market, whereas nothing on the Internet is fixed. The plan needs to incorporate a slow transition from the RichWin business until the Internet begins to pay the bills."

Zhidong agrees, though he is confident that there are strong sources of revenue for an online business. Some examples: Web exclusives on events such as the Olympics and the World Cup could generate enormous advertising revenue. Online and borderless commerce among the world's Chinese will be immensely profitable, too. All of Zhidong's plans are dependent on a global Chinese audience. Foreign Chinese often have family in the Mainland. Not only can they keep in touch online via e-mail, but they can easily send flowers and presents or share photographs and videos online. The foreign Chinese market will help support the Mainland market at the beginning, Zhidong says. Foreign Chinese are generally more affluent. For that reason, the potential for advertising abroad is greater than on the Mainland. In addition to having more spendable income, foreign Chinese are more sophisticated shoppers and have credit cards, which is essential for high-volume Net commerce. On the other hand, the nascent China market is expanding.

"Software companies are safer," Bo interjects. "Once you get into general content, you're a media company. A global website is going to run into the problems of local markets plus local governments."

That's why Zhidong sees a network that is both "country specific and global." Some offerings on the websites will be different in China and Taiwan. News, which must be acceptable to both governments, is the obvious example. However, content about sports, travel, business, entertainment, and the like is appropriate for a worldwide audience.

"Fine," says Bo at last. "If the premise of the network is that it's global, do you plan to try to launch SRSnet in the U.S., Taiwan, and Hong Kong?"

Zhidong says, "That's the problem. We don't have expertise outside China's borders. I see three choices. One is recruiting a team and setting up a small office in Silicon Valley. Two is allying with a leading player—a Netscape or Yahoo—to become their Mainland partner. Third is an acquisition or merger with a portal covering the world's Chinese outside China."

Near the bottom of the wine bottle, the conversation about SRS wanes. Looking out at the view and feeling lightly tipsy, they reflect on the unexpected course of their lives.

Zhidong stretches his thick arms and loosens the collar around his neck. "When I used to work on the computer, it never occurred to me that I would be a businessman," he says. "Sometimes I wish I could just write programs."

Bo sighs. "We were taught in my family that business was beneath us," he says. "Merchants were crass, interested in money. My father valued art and music and philosophy. Ideas."

Zhidong says, "We were never like that. We wanted to make some money. It all depends how you do it. There is an honorable way and a dishonorable way."

On the following day, Zhidong meets with another friend and adviser named Daniel Mao, who is a savvy venture capitalist with Walden International, one of the investors in SRS. (Daniel is on the SRS board.) Mao, a slender and eager man with a round face and thinning hair, has agreed to help Zhidong investigate the options. He researches the existing Chinese portals. In addition, he sets up meetings at Yahoo and other Internet companies so that Zhidong can discuss the possibility of a partnership. He also introduces Zhidong to a series of engineers and company chiefs throughout the Silicon Valley who provide different perspectives on the technical, economic, and political problems associated with a global network.

The conversations continue next in Beijing at a Zhongguancun teahouse with Daniel, Yan Yanchou, and Wang Yan. Zhidong explains the issues as he sees them, including the three options. "There are problems inherent in them all," he says. "We'd have a tough time going global. Chinese communities in the U.S., never mind Taiwan and Hong Kong, are distinct. In addition, building teams in the various markets may be beyond our skills and bandwidth. The problem with working with one of the big Internet companies is that we'll be marginalized—one of hundreds of partners. The other choice is a merger or acquisition. Are we ready for that? If we grow too quickly, we lose the ability to plan and manage growth."

Everyone agrees that time is critical, so Zhidong's top advisers are charged with determining the best way to proceed. Daniel Mao recommends a meeting as soon as possible with the founders of a portal called Sinanet. Daniel knows and respects them, and says that there is synergy with SRS both in terms of geography—Sinanet covers the Chinese American and Taiwanese communities—and in the personalities of the teams.

Sinanet's origins are archetypal Silicon Valley: three Stanford students, a cramped living room, and a shared, restless idealism. One of the three, Jack Hong, was born in Taiwan and grew up in Ohio, where his father was a professor. His family moved back to Taiwan when he was ten, then returned to Ohio when he was fifteen. "There was a Korean in my twelfth-grade class," Hong says. "Everyone thought he was my brother." Though he had never been to Mainland China, he felt drawn to his Chinese heritage, partly because he felt like an outsider in both Taiwan and the United States. In college, at the University of Texas in Houston, he founded a band that played Chinese music. He started another Chinese band while attending graduate school at Stanford.

Studying art, product design, and mechanical engineering, Hong tutored on the side. One of his students was Ben Tsiang, who came from a prominent Taiwanese family. (Tsiang's grandfather was Chiang Kai-shek's chief of staff.) Tsiang had his sights on the Harvard Graduate School of Design.

Before graduating, Hong also met Hurst Lin, who was born in Taiwan and grew up in New York before earning his Stanford M.B.A. After Lin graduated, he worked as a financial consultant while Hong stayed on at Stanford for his Ph.D. At night, he'd invite Lin and Tsiang over, and the three would sit around Hong's living room discussing their futures. They decided they would work together in some business and considered several ideas.

Lin and Hong attended a presentation by Jim Sha, one of Netscape's founders, in 1994. Sha was recruiting at Stanford. He gave the audience a preview of things to come. "He was showing these hockey-stick charts of the growth of the Internet, and we thought, 'Yeah, right,' " Lin says.

Soon after they missed the Netscape boat, Lin says he began to "get it." Hong, too, got online and into the Net's implications for distance learning; he wanted to create a site devoted to online education. Then Tsiang came up with the idea of a Chinese site.

"We didn't have a competitive advantage in education services," Lin says. "But we had one in a site for and by Chinese. We were Chinese. We knew the needs of Chinese Americans."

The more they discussed it, the more they became inspired. "Chinese kids in the States lose their Chinese, lose touch with their roots. I thought the Net might be a hip place for kids to read in Chinese, as opposed to boring newspapers," Lin says. "Also, there was some chauvinism thrown in: Why should the computer world be dominated by English?"

In April 1995, they made their move. For their company's name, they rejected Great Wall ("Too trashy," says Lin) and China.com (it didn't include Taiwanese, Singaporean, and other Chinese people). (Another company later took the name China.com.) They settled on Sinanet and, courtesy of hijacked Stanford computers, launched a site. Hong used his art and engineering skills to create software that rendered Chinese characters on the Web. For content, the new partners called Chinese embassies and consulates, both of which receive a daily news feed from China's State News Agency. "Sure," Lin admits, some of it "was official news and propaganda. But we had to put up something."

Every morning at six o'clock California time, one of the Sinanet founders uploaded the news from New York via a 14.4 modem. It took about half an hour—more if the modem disconnected and they had to start over. They sought other news sources and hit pay dirt when the second-largest Chinese-language newspaper published in the United States agreed to let Sinanet have its content for free. (The publisher viewed it as good publicity.)

"Before us, news from China was always really late—sixteen hours late if you lived in San Francisco, where there is a large Chinese population and some Chinese newspapers, and as much as three days late in other parts of the country," says Lin. "We were able to beat the newspapers. We became a big hit among Chinese students everywhere outside of the Mainland." Hong says serious news from China was the most sought after, followed by entertainment and cultural news. There was also a hearty audience for tabloid stories—grisly murders and sex scandals. "Like the time when a bunch of kids got chopped up," he says.

Sinanet put up chat rooms and bulletin boards, too, since they were "self-generating content," as Hong says. Users from Chinese communities on college campuses spread the word to their families, and Sinanet's traffic steadily grew. To pay for the site, they began selling Chinese food online. "At least it solved our lunch problems," says Hong. Stan Saih, the founder of Acer Computer, heard about the site and offered operating funds and servers and computers—which became particularly useful once Stanford kicked Sinanet off its system.

As Sinanet's traffic grew, the founders settled into a real office and wooed a CEO, Daniel Chiang, then president of Trend Micro, an Internet virus protection and security software company. With Chiang, a towering former Taiwanese basketball star, in charge, the three founders took VP roles and focused on supplying editorial content through deals with two hundred providers, beefing up the arsenal of servers, and successfully selling advertising to a variety of companies interested in reaching the Chinese American community.

There was, however, a missing piece to their plan. As Hong deadpans, "It was just a small missing piece, but a missing piece nonetheless. It was China." And, Lin adds, "As a bunch of badass Chinese Americans, we could see pretty clearly that there was no way we could parachute into China."

Sinanet's trio of founders and new CEO were researching a move into China at the same time Wang Zhidong's SRS team was exploring ways to expand beyond the Mainland. At their first meeting in the fall of 1998, both sides immediately saw the logic of a merger.

The initial meeting with Wang Zhidong and Wang Yan and the three Sinanet founders plus Daniel Chiang goes well, though both camps are irresolute and protective of their respective companies. More meetings lead Wang Yan to be inclined to move ahead. "There is a common willingness to be biggest in the world," he says. "Together we decided not to compete but to cooperate." On Sinanet's side, Jack Hong says, "The synergy is obvious. Our technical teams respect one another. We have a purpose in common."

Zhidong still has doubts, however, and the decision to go forward is preceded by months of soul-searching, all-night tête-à-têtes with friends, family, and advisers, and numerous transpacific flights and conference calls. In November, he sequesters a group of his closet advisers in the Dragon Club at the Longyuan Hotel in Beijing. There is no sleep for forty hours. Finally, he's ready. The promise of reaching tens and someday hundreds of millions of the world's Chinese proves too great to pass up. Sina.com, as the new multinational will be called, has the potential to connect Chinese speakers worldwide and, as such, unite them as never before socially, culturally, and commercially.

Negotiations begin in late November. It takes thirty-two days and almost as many sleepless nights to iron out the details of the newly formed East-West alliance. The much larger company—SRS has profits, never mind revenues that are far greater than Sinanet's—is essentially acquiring Sinanet, though there are aspects of a merger in the deal. For now, Zhidong remains CEO and president, Daniel Chiang is vice CEO, and Wang Yan is general manager of China Operations for the website. Yan Yanchou remains the chief technologist in China. In Sunnyvale, the U.S. president is Jack Hong. Daniel Mao, enticed to leave Walden to join up, becomes the chief operating officer.

Wang Yan leads an ad hoc team—made up of staff from both companies—in an effort to combine the sites so the look and feel is the same whether one logs on from the United States and Europe (www.sina.com), China (www.sina.com.cn), or Taiwan (www.sina.com.tw). (Hong Kong [www.sina.com.hk] is added later.) Though the design is integrated, as Zhidong envisioned, political and cultural differences dictate that the sites remain unique. Local editors in each country are in charge of selecting content relevant to (and legal in) their locale and supplementing it with articles on local affairs. Some choices are obvious—entertainment listings, for example—but the Chinese editors have to make others as well, scouting for content that won't offend government censors. The initial political caution later gives way to news and discussions that are relatively unfettered and uninhibited, but content remains a precarious balancing act. While far less used (and at the time less respected by Chinese) competitors, including the site that takes the name China.com, kowtow to the powers that be, posting only preapproved news, Sina.com makes its own selections and pushes the limits when its editors deem it to be appropriate.

Now that the new company is official, Zhidong's dream is crystallizing. Given the virtual Great Wall that has surrounded Mainland China, blocking all but government-approved information and communication to and from China's billion-plus people, Sina.com feels like a miracle.

In the West, the Internet is surging throughout the late 1990s along with the business behind the Internet. Since the 1995 Netscape initial public offering, the Net has lead an unprecedented economic boom. Companies that didn't exist a year before are valued at multiple times their bricks-and-mortar counterparts. Pure Net plays—Yahoo is the best example—seem unstoppable, with stock prices and valuations in the stratosphere. Dot-coms and other Internet start-ups are like weeds in the Silicon Valley and other U.S. and technology centers. There are a few prognosticators of calamity—*It's a bubble! Bubbles burst!*—but they are roundly dismissed as bores and bad sports.

It's no wonder that part of the immediate Sina business plan is a public stock offering. It's not just about participating in the gold rush, though making millions wouldn't be bad as far as the principals of the companies and their investors are concerned. There's also a business imperative to go public in order to compete in the Internet space. Sky-high valuations put newly public companies in strong positions to hire the best talent, invest in R&D, and make strategic acquisitions. Since Sina has the irresistible combination of the Internet and the unexploited market of the largest population on Earth, when the Sina.com offering is announced, investment companies rate its as red hot.

Positioning itself for the IPO takes shrewd planning, however. The heads of the new company have already agreed that a multinational portal needs a strong, experienced leader. Zhidong is nervous about the decision to bring in someone from the outside—hiring his own boss!—but he wants to do what's right for Sina. It's not a tactic for the IPO. Zhidong feels that Sina will benefit from a manager who has run a large company with branches across the globe. It's a risky but logical move, and Zhidong approves a CEO search, convinced that his position as founder can be protected even if a professional, experienced manager is brought in.

The most prominent candidate is Jim Sha, the man who ironically had given Sinanet's founders their first look at the Net back in 1994. Sha, who was born and raised in Taipei, is the son of famous Taiwanese moviemaker Yung Fung Sha, whose *Touch of Zen* won a Best Picture award at Cannes in 1975. After studying electrical engineering at the prestigious National Taiwan University and obtaining a graduate degree at UC-Berkeley, Sha worked at a number of Silicon Valley companies, including Intel, where he built a 386 emulator. It was a few years later, when he was at Oracle, that a recruiter called and said he ought to "take a few minutes to meet this guy Jim Clark." Sha and Clark, the founders of Silicon Graphics, met at Buck's diner in Woodside and hit it off. Clark convinced Sha to join his new company, Mosaic, as a consultant. He next followed Clark and cofounder Marc Andreessen to Netscape, where he ran R&D and managed two spin-off companies. He retired just before Netscape was sold to America Online. "I'd had enough rides on the Internet roller coaster to last a lifetime," he says. He planned to manage an investment fund with other Netscape alums.

When he hears about Sina.com, however, he agrees to meet Daniel Chiang, Daniel Mao, and other representatives of the company. He warns them about his resolve to retire, but they convince him to listen to Zhidong. When the two men meet, they talk for hours. Sha is moved by Zhidong's vision of Sina. Afterward, Sha says that he would have turned down any offers that came along except for this one. "The call of China was too much to refuse," he says. "I joined against my wife's wishes. I felt obligated."

It doesn't take long for Sha to begin to transform Sina. He streamlines the new organization, launches the Sina mall, and brings in big-league marketers and financial consultants, all the while setting the stage for an IPO. It doesn't hurt that in a survey conducted by China's Internet authority, China Internet Network Information Center, a branch of the MII, the new site is voted number one by China's Internet users.

The buzz in the investment world grows. Sina's numbers keep improving, and advertisers embrace the global site. Sina Mall begins to rack up tentatively impressive sales. As Michelle Yeh, the Sina marketing director brought on by Sha, says, "We find that people come to us to send flowers to their mother in Taiwan before they go to 1-800-Flowers, because they trust that we will do the job better in a country we are part of." Visitors to Sina outside China are using the site to buy hard-to-come-by products from home, such as Moon Cakes for the annual mid-autumn Moon Festival. Sina begins to generate impressive advertising revenue, too. Where else can advertisers go to reach Chinese users around the world? One Sina offering that is incredibly popular is Club Yuan. Yuan means "destiny." Club Yuan is Sina's dating service. The company doesn't have to look far for testimonials: Three Sina managers meet their wives through the service.

There is a lot to celebrate. "We know that the ideological and cultural problems are bigger than any technical problems between China and the rest of the world," says Jack Hong. "First Amendment? China doesn't have a First Amendment. Yet what brings change but openness—access to information and the ability to communicate? It's happening at Sina and it's all very gutsy and exciting. It's like a spiritual awakening."

"Suddenly, we are no longer separated by geography and nationality," Zhidong observes. "Finally we are genuinely tied together by the Internet. Now Chinese in every part of the world are able to communicate. Sina is becoming a digital living platform for all Chinese people around the world."

It's not seamless, however. Over the course of the next couple months, Zhidong begins noticing worrying signs. It's internal, subtle. He quietly observes that Sha, who originally embraced and shared Zhidong's vision about the paramount importance of the Mainland audience, is emphasizing the non-China market at the expense of the nascent, but growing, one in China. "If we gain the world but lose China, what would be the point of it all?" Zhidong says to Bo one chilly spring morning. It's the first time Zhidong admits aloud that he's troubled. He says, "I hope I won't live to regret hiring Jim."

In the ensuing weeks, Zhidong watches as Sha brings in more and more of his own team and gives them power over Zhidong's group in China. He worries that Sha is lessening the influence of Yan Yanchou's crack technical team in favor of his own engineers in Sunnyvale.

When I speak to Sha, he tells me that he has great respect for Zhidong and views him as a friend, but he was hired to lead Sina and he must pursue the course he deems best. Zhidong, however, is feeling shut out of decisions. His team in Beijing is increasingly sensing that they are being condescended to by their U.S. counterparts. More top people that are brought in are installed in California, not Beijing, where it is beginning to seem not like the center of the Sina network, but an outpost. Slowly at first and then with increasing anxiety, Zhidong worries that one of his worst fears is coming true. Someone—someone he had hired—is attempting to take control of his company. He worries that the current course could prove disastrous. If Sina under Jim Sha ignores the Mainland, China will once again be the stepchild. Zhidong's commitment to expand globally was predicated on the understanding that the center of the Chinese world is China. He maintains that a global network that minimizes China will be a commercial mistake. Worse, it will ignore—in fact, defeat—his vision.

It leads to more sleepless nights. He feels as if he has to do damage control after many of Sha's decisions. At the same time, he has to operate covertly, since Sha is the company's chief executive.

In this climate—of exhaustion, worry, growing anxiety—on May 7, 1999, at the hour that is inhabited by insomniacs and vampires, Zhidong plays out disaster scenarios in his mind. He cannot believe that he is living a high-tech version of a palace coup, with him as the victim. Neither can he ignore a long list of other worries—the kind that come with any high-tech start-up: server crashes, scaling quickly enough, myriad other technical problems. In the months following the merger that created Sina, Zhidong has been confident, but mergers are messy, and Sina's, even after the hiring of Sha, has continued to involve a high-wire act of integrating the Chinese and U.S. executive teams, managing vastly different corporate cultures eight thousand miles apart, and navigating the unpredictable and amorphous regulations of various Chinese government agencies. He knew that the alliance could bring challenges to his leadership, but he didn't expect that his fears of being marginalized would be realized so quickly. In the meantime, he is trying to manage the internal debates that rage about how best to position the company for its upcoming IPO. There is no end to delicate diplomatic challenges, such as placating managers on both sides of the Pacific so they don't resent one another. Plus a long list of technical, operational, and personnel issues that come with any Internet venture, particularly with all of the salivating competitors. Oh, yes. And there's the added uncertainty of China! On the one hand, the unfathomable potential. A revolution of communication and information. What will shape up to be—what *is* shaping up to be the world's biggest Internet market. As remarkable an opportunity to present itself in generations. On the other, complete political and social uncertainty. Yet even after all this—the risks of any merger of this size, the volatility of the Net and the Net economy, plus the added risk of China—Zhidong is unprepared when, at 6 A.M. China time, American jets bomb the Chinese embassy in Belgrade, Yugoslavia, killing four Chinese staffers and maiming two dozen others.

Yan Yanchou is in Shanghai for a series of meetings. One lasts all night, and he returns to his hotel room in the early morning. Before catching a few hours' sleep, Yanchou, whose sleepy manner belies his reputation as one of the heroes of the software industry in China, turns on his laptop and logs onto the Sina website, where more than 10 million Chinese people around the world now regularly turn for news and to communicate with one another. Posted on one of the site's bulletin boards is an urgent message from a user in Yugoslavia. It says that ten minutes ago, NATO dropped a bomb on the Chinese embassy in Belgrade.

Yanchou reads and rereads the dispatch and then searches Sina.com's news services for more information. There's nothing. Could it be a twisted joke? A dozen or so minutes later comes the first official news report from a French news agency confirming the bombing. Yanchou places a call to Wang Zhidong in Beijing.

As usual, first thing in the morning Zhidong tries to log onto Sina.com to check his e-mail and read the day's news, but he is unable to get a connection. That's when Yanchou's call comes in. When he hears the news, Zhidong knows why he can't reach the network and predicts that the rush of traffic won't let up. He instructs Yanchou to switch every Sina server over to cover this news and arrange to buy more servers and get them up and running as quickly as possible. Zhidong next heads across town to the office as Yanchou jumps on the phone to his engineering heads before flagging down a taxi for the airport to catch the next flight to Beijing.

By midday, both men are at the Sina offices in Zhongguancun, where they watch the latest traffic report: There are 8 million page views, already four times a typical full day's traffic and millions more than the previous record (a chat with a famous anchorman), before the counters are no longer able to track the volume. Throughout the day, Sina.com's users throughout China, plus those in Europe, the United States, Hong Kong, and Taiwan, many of whom are convinced that the attack was intentional or even an act of war, devour news reports and debate in chat rooms and on bulletin boards. The engineers reset the servers and three minutes later the limits are again reached.

At Sina, fans whir. A staticky radio on a metal desk in a cubicle plays opera. In one engineering room, a couple dozen engineers in open collars and T-shirts work in front of computer monitors, feverishly clicking keys and the buttons on mice. Down the freshly whitewashed hall, in his small office, Zhidong is monitoring usage, which is still increasing. The new servers are configured and up and running, but they still intermittently crash. The engineers down the hall reset them and switch them back online. Boom. Another crash. It will be a long night.

Across town, Bo's taxi drops him off at Shunya Restaurant, where Edward is waiting. The two men exchange handclasps. Edward has just returned to China from California, where he was at the birth of his and Jean's second child, a girl named Frances. At a table they drink down beers and discuss the still-fresh rumors about the bombing—that it was a planned attack on China for secretly helping the Serbian army; that the leaders in Beijing are calculating the most effective response; that prominent Chinese are arranging for their children, abroad in American universities, to fly home immediately. Like almost everyone in China, the news preoccupied Edward and Bo's days. They tried to work, but it was impossible to get much done. Bo first spoke to Edward in the morning and then again after noon, when they agreed to meet for dinner. They gravitated together because of their great friendship and a fact common to both of their lives. Just as Bo has Heidi and Tiger in San Francisco, Edward has Jean, Stephanie, and the newborn baby in Dallas. Their families are in enemy territory, at least according to many of their friends, friends who should know better.

Because of the wide ocean that divides their families, it is impossible to separate the tension between China and America and the tension in their lives. The bombing exacerbates the problem, but it's a constant. It goes back to my conversation with Bo on the train. His and Edward's values, their lifestyles and their self-images in China and in the United States, are in conflict. So are the demands and expectations on them. To keep a life together in either country is tough enough. To keep a family together anywhere is, too. To keep them together in two cultures, and particularly these two inherently contradictory cultures, often feels impossible. Now what? If the enmity between the two countries escalates, if the mounting distrust between the Chinese and American people worsens, if it leads to a conflict—there are calls for a counterattack from people who a week ago were praising President Clinton!—what will they do? Their lives would be sliced in half.

"What would you do if there was a major schism—if relations were cut off?" Edward asks Bo.

"I don't know. Heidi loves living in China, but both China and San Francisco are our homes. We want our children to grow up knowing both of our families and both cultures."

Edward sips his beer. "Cooler heads will prevail," he says. "For now, there's nothing we can do."

Bo says, "China and the U.S. will be closer in the future, not further apart. We have much to offer each other."

Edward nods soberly. "You and I cannot change a lot of things on a political level," he says, "but our work can make it less likely that divisions like this will fester." Edward's sincerity touches Bo. "When people know each other," he continues, "when they talk every day and have friendships, there are fewer opportunities for misunderstanding."

"I don't know," says Bo. "I hope you're right."

They talk until late at night, past when the restaurant's other patrons have departed. The staff mills about nearby, waiting, checking out the reports on the tiny television that is switched on in the kitchen. At the table, Bo telephones Zhidong at Sina to check in again.

"We're just trying to keep the site up," Zhidong says. "What a time to have signed a Sino-U.S. joint-venture deal." Indeed, if things degenerate between China and the United States, Sina will be split in two, as well. In Sina chat rooms, there are predictions of World War III. Nevertheless, Zhidong says that he hasn't had much time to worry about the fact that he has just joined forces with an American company. "A different truth is hitting," he tells Bo. "Think about what we're seeing on the site. People in China are reading real-time news from Belgrade, talking to one another whether they are in Beijing or Paris or New York. It has never happened before this."

Bo and Edward leave the restaurant and walk through the bamboo gate that leads to a path along the Zha River, which cuts through central Beijing. A bicyclist wearing nothing but short pants pedals by. The din of his portable radio can be heard: news from Bosnia. Zhidong has reminded them of the larger importance of their work, but there's nothing to be done tonight. They are stuck with the terrifying fragility of their lives.

Even as America offers its lame excuses (old maps!), millions of Chinese inside and outside of China continue to gather online to read the news and console one another. It's indisputable proof that Zhidong is correct: The Internet is already beginning to succeed where internal and foreign protests, finger wagging, summits, and threats of sanctions have thus far failed. The Net is opening up China. Any doubts are put to rest in the aftermath of the bombing.

Following the mind-boggling blunder, justifiably outraged Chinese citizens take to the streets of Shanghai, Beijing, and other cities for government-sanctioned—maybe even government-organized—protests against the Yankees. Marching within a mile of Sina, Beijing University students rail against the United States, while online they post sentiments like "Pay blood debts in blood." There is no letup in the volume and volatility of conversation on the network, which proves once more that the less-than-auspicious week for U.S.-Chinese relations is indeed historic for the Internet in China. This is the first time that ordinary Chinese, albeit privileged ones with access to the Net, are able to participate—uncensored—in a global event. Originally Zhidong was drawn to computers as a way to express what had been hidden inside himself. Now his network extends the ability to millions of people in China. He has an image of people sitting in front of their computers and, through the machines, touching one another whether they are in Beijing, Taipei, Paris, or a remote village in the north of China. Yes, it's a precarious time. There is political volatility along with the internal problems at Sina. However, the power of the Net itself and Sina in particular is undeniable. Zhidong hasn't much time to dwell on it, but he allows himself a fleeting thought. This is your dream: uniting Chinese people everywhere.

In China a month or so after the bombing, I find a different climate entirely. It's night and day from six months ago. My friends don't act differently, but many Chinese do. On my previous visits, there had been a generous and enthusiastic welcome added to the general friendliness specifically because I was American. Some of it had to do with history. Some older Chinese recalled the U.S. soldiers who liberated them from Japan's invasion. Since then, the people of China have become increasingly interested in American ideals—ideals they learned about because of the opening economic and social ties between our two nations. There was a wariness among some that China could become too Americanized, but in general there was an open affection toward the United States. One thing that helped was the recent visit of President Clinton in June 1998. A conversation between Clinton and President Jiang Zemin had been broadcast live on Chinese television. Hundreds of millions of people were astounded by Clinton. Apparently the Chinese had never seen anything like him. They seemed awed by his spontaneity and passion, particularly compared to Jiang, who seemed inimical and icy. The contrast embarrassed many Chinese. I heard often about how much the people liked and respected Clinton. Americans already seemed to be in general favor, and the Clinton visit added a new wave of respect. However, the bombing seems to have changed that. The general amiability has been replaced by suspicion. It's palpable in front of the U.S. embassy, where protesters are camped out. One holds a poster of a bomb falling on its target, the Statue of Liberty, in poignant contrast with the copy of the Liberty statue erected by the student protesters in Tiananmen Square in 1989. Since the bombing, the Chinese seem angry with America. In gatherings, I find myself on the defensive. I hear, "Americans have the gall to discuss China's human rights, yet look at the way black people are treated in your country. How many of your black citizens are in jail? Look at the violence. Alienated children shoot their schoolmates. Anyone who claims that the U.S. isn't an imperialistic aggressor is a dupe. You decry the shootings at Tiananmen Square, but your government killed your own students at Kent State. You annihilated the Native Americans and anyone who stands in your path of so-called progress."

Bo, Edward, and the others whose lives are divided between the two countries feel torn. Edward, however, finds a heightened sense of purpose. "The more contact, the more communication, the more relationships—electronic or otherwise—the less likely for an escalation of conflict and tension," he says when we meet at AsiaInfo's office. "The bombing was a reminder to anyone who might have forgotten the gravity of our mission."

The mission. It's what they hold on to. Zhidong's mission is intensified, too, even as there are new crackdowns on the Net. On the ten-year anniversary of Tiananmen Square—June 1999—Internet cafés are once again closed down throughout China. Nonetheless, there is another spike of discussion, including angry clashes of opinion between supporters and critics of the CCP, on Sina and other websites. More surprising, there are thousands of entries in the *People's Daily* chat room. The *People's Daily!* The Chinese Communist Party newspaper! While it's not surprising that anti-U.S. sentiment is allowed—"The U.S. is a killer—we can't forget [the embassy bombing on] May 8"—what's surprising is that some of the conversation is harshly critical of the Chinese government's attack on the students in 1989. A report in the *New York Times* discusses the dialogue on the *People's Daily* site that precedes a visit by a Western journalist to China. "I want her to know that we Chinese people don't like our government," writes one contributor. Another says, "The Communist Party doesn't let Chinese people read newspapers from overseas!" Jiang Yaping, who heads the Internet division of the *People's Daily,* tells the *Times,* "We try to let the people speak. We are the main newspaper of the party, so of course there are limits. But it would take something stronger than these to be [deleted]."

It's also stunning to note that the government itself is using the Net for things that sound suspiciously . . . well, democratic. Of course some of the more than three thousand official government sites peddle Maoist and other Communist propaganda, but in the summer of 1999 the State Development Planning Commission opens a website on which it solicits views from Chinese citizens about their hopes and expectations for the country over the next five years. Thousands of people respond, asking for everything from a serious effort to eliminate corruption to more spending on education. In an expression of previously unthinkable dissent, one respondent writes that the Net survey is "a waste of time until the government grants Chinese real freedom of speech." Whether the government actually listens is another story, but Chinese people tell me that this is the first time they have ever been asked for their opinion.

CHAPTER 10

THE eCHINA SYNDROME

In November 1998, Bo meets with ChinaVest's Jenny Theleen at the Swissôtel bar. Over the past months spent searching for entrepreneurs, Bo has selected a few with particular promise. He worked with them to polish their business plans and then recommended them to his bosses. ChinaVest declined to invest. In fact, Bo complains that he hasn't even heard back about his most recent proposals. Tsu advises patience. "It's a matter of time," she says. In the meantime, she asks if he will help out with the first Beijing location of one of the companies in ChinaVest's porfolio, T.G.I. Friday's, the restaurant chain. "Will you go there and spend some time? See why it's not taking off. Maybe you have ideas that could help."

Bo is piqued. "T.G.I. Friday's doesn't work in Beijing because it's not Chinese," he says. "It's a stupid idea in America and a stupider idea in China."

After the meeting, he says, "They say they want to invest in the Internet, but they are used to investing in companies with traditional business models. In the West, future growth and the assumption of future profitability is taken into account. They say they want to be part of the new economy, but they don't invest in it."

Bo does not go to T.G.I. Friday's. Instead, he spends much of his time at AsiaInfo and SRS, helping Edward and Zhidong pilot their companies forward. It's far more gratifying for him, though his bosses at ChinaVest harangue, "Are you working for them or for us?"

Bo says, "I left investment banking to become a venture capitalist so I could find and then work with companies. If I can't do that, what am I doing?"

ChinaVest may turn out to be right in their cautious approach to technology (in hindsight, of course, they were, and the company has since increased its portfolio with a series of sound investments in the sector), but it's maddening for Bo, who is exceedingly impatient. Sandy Robertson allowed Bo free reign, whereas the ChinaVest managers increasingly keep tabs on him. "When I'm running out to a meeting, they ask where I'm going!" Bo says. He sighs. "The last thing China needs is imported fast-food chains. Most Chinese people don't even know what T.G.I. Friday means."

Bo is home in San Francisco and wants to spend all of his nonworking time with Tiger. (Back in China, Gong Hongjia jokes that Tiger Feng is "the most successful Sino-American joint venture so far.") Bo and the baby are on similar time schedules—awake at odd hours throughout the night. Lihui, who bustles around the house in her padded jacket and silk pants, is overjoyed to be able to be a doting grandmother. She and Bo do predawn baby duty and Bo works the telephone, calling China. He sleeps a few hours and scoots off to meetings during the day. Between them, he makes pit stops at home and takes Tiger to the playground or cafés. After growing up in the Cultural Revolution, Bo's demonstrative and loving attention to Tiger is not what one would have predicted, since people are said to replicate their childhood when they become parents. It's inconceivable to him that he will soon be on a plane that will take him six thousand miles away from his son.

Before departing for China, Bo visits Sandy Robertson at his new office. Sandy's life has been anything but staid. Less than a year after Bank of America took over Robertson Stephens, BA was part of a controversial $133 billion merger with NationsBank. Since the resulting company had two investment banks, Montgomery Securities as well as Robertson Stephens, one was put on the block to be sold; RS & Co. was acquired by BankBoston. Sandy didn't go with it; after twenty years, he left the company he founded. He is currently based in a friend's office that is also in the Bank of America building, from where he is managing his own investments while considering the formation of a new company when, in 1999, his noncompete agreement is up.

Robertson, always gracious and impeccably dressed, greets Bo. Outside the window, the Oz-like San Francisco skyline seems painted.

Bo has already confided to Sandy about his disillusionment with ChinaVest, but hasn't broached his latest idea. He boldly asks for Sandy's opinion about a new company, a hybrid between an investment bank and a VC fund, what he describes as an advisory business that will provide a range of services to start-ups—fund-raising included. "It will start out brokering deals," Bo says, "though probably smaller deals than AsiaInfo and SRS. Backing the type of companies that I have seen but have been unable to finance over the past year: early stage companies with enormous promise. For a fee and equity, we will provide a full range of advisory and business services that will allow them to focus on their technology. The goal will be to eventually raise our own fund, but at first we'll bank deals."

It doesn't take long for Sandy to make good on his promise to work with Bo again. Robertson enthusiastically agrees to back the company. They hash out the details and agree on the name Robertson, Feng Technology Associates. To launch RFTA, Sandy will put in an initial investment of $400,000. Sandy's name means even more than his investment. It's a testament to his commitment to the idea and to Bo.

Bo finds it difficult to respond, but he says, "Thank you, Sandy." After a moment, he asks, "Are you sure?"

Bo knows this is an extraordinary act of faith on Robertson's part. Bo has seen that any involvement by Robertson brings credibility and opens doors. With this degree of commitment from Sandy, Bo's stature in the investment world will grow. At the same time, however, the new company's capital is limited. It's a start-up, not an established company like ChinaVest. He's nervous, but when he leaves the meeting, Bo is elated. Elated and terrified.

In Beijing, Bo meets with Jenny Theleen and resigns. The world of the China Internet is small enough that it's best if you can keep good relationships with the main players. In spite of experiences that both sides would describe as negative, Bo and the ChinaVest leaders know that they will likely work together in the future. The meeting concludes with mutual good wishes.

Within the week, a handsome logo is created for the new company and the name is emblazoned on briefcases and baseball caps. Bo rents a small office in Shanghai. He chooses Shanghai rather than Beijing for a number of reasons. Beijing has Zhongguancun, but Shanghai has a growing silicon valley of its own. (There are also growing technology centers in Shenzhen and other cities.) Whereas Beijing has Beijing University and such companies as SRS, Sina, and Legend, Shanghai is China's finance center and has a growing number of start-ups. In addition, Shanghai has been one of the most aggressive cities when it comes to Internet investment and innovation. A side benefit of Shanghai is that the central government seems somewhat less oppressive. There's a Chinese saying about the breezier atmosphere of Shanghai, "Tiangao huangdi yuan," which means, "The sky is high and the emperor is far away." In spite of all that, Bo probably would have settled in Beijing if it weren't for Heidi, who is spending more time in China so the family can be together. Shanghai is an easier city for families because of an abundance of new parks, milder weather, cleaner air, and less traffic. Bo's brother Tao and his family have also moved from Beijing to Shanghai. Lihui, happy to spend time in her hometown and have all of her grandchildren close by, visits often.

Bo rents a modest apartment not far from the Kerry Center. He can walk to work in twenty minutes. There's a *longtang* that cuts through several brick apartment complexes and brings Bo past one of his favorite tea-and-dumpling shops, where he often stops for steamed buns and pork-stuffed wonton in peppery broth. From there, he cuts across a couple of side streets and he's at the new RFTA office.

It's a small suite with teak furniture and a view of the bustling Nanjing Road. There are three desks—Bo's, a secretary's, and Bo's first employee's. Jiang Shaoqing introduces himself to me as Jeffrey, explaining that he chose the American name during a high school English class because it reminded him of a British prime minister. Before RFTA, Jiang was working for the Sakura Bank when, as he puts it, "Bo conquered me." The story reminds me of when Steve Jobs convinced John Sculley to leave PepsiCo to become CEO of Apple Computer. "Do you want to continue selling sugar water to children or change the world?" Jobs asked. Bo told Jiang to come with him and work "to change your nation, to build a greater world for your children and your children's children." Jeffrey says, "I didn't have children, but you can't say no to Bo."

Jiang's title of analyst hardly describes his range of jobs at RFTA. (Hardly anyone at a Chinese start-up does what his job title implies.) Jeffrey does whatever Bo needs—greets visitors, translates, analyzes the books of potential clients. After three months in his new position, Jiang speaks of Bo with reverence, addressing him as Feng Xiansheng, or Mr. Feng, the formal way. "Convinced by Feng Xiansheng, I am very excited to come to this company," he says. "I want to do something for my country. It's time for change and we have the power to build a new type of China—modern and prosperous and free."

Bo's staff grows larger. After Jeffrey, he hires another analyst and more office help, all squeezing into the same suite. Bo checks in daily, but he rarely occupies his desk. His days are nearly identical to when he was with ChinaVest and Robertson Stephens before that. That is, he's on the move. He meets entrepreneurs. He visits companies. He meets potential investors. He's rarely in any one city for more than a few days at a time. He meets more entrepreneurs. He visits more companies. By now Bo has amassed enough frequent flier miles to fly for free to anywhere on the planet, though free flights are the last thing he wants. It's like awarding a deep-dish berry pie to the stuffed-to-the-ears champion of a pie-eating contest.

On an early morning in Shanghai that is pristine clear—a metallic blue sky, puffs of cumulus clouds—I board a China Eastern flight to

Beijing to join Bo and Jeffrey, who are working on a series of deals. Jeffrey greets me at the airport terminal in a rented Santana. Behind the wheel, he looks smaller and younger, barely old enough to drive. (He's twenty-four). He has a thin face and distinctly dark features against the opaque complexion of someone who works all night and rarely sees the outdoors. He wears a cream blue suit even though Bo announced that the season calls for "summer casual." (As usual, it's scalding.)

The VW glides on the freeway past *yang* trees, like fat-leafed aspens, until we slow down at the outskirts of Beijing, with its smoggy-brown turbid sky. The car, with black curtains and lime slipcovered seats, has to stop while two men, one with a shoulder pole carrying water and another with a wheelbarrow filled with an enormous dead pig, cross the street. Traffic is dense with the usual swarm made worse by a collision between a bicycle rickshaw and an enormous van painted with a chartreuse fire-breathing dragon. "Enjoy life and smoke more," squeals a robotic voice from the van's trumpet-shaped speakers. The incessant pitch mixes with the screaming of the rickshaw's driver, honking horns, a siren, and the ringing of Jeffrey's cell phone programmed to play a chirpy version of the Chinese national anthem.

Jeffrey and I finally arrive in the Zhongguancun, where we park in front of a somewhat seedy multistoried building. The lobby, with windows covered with a metal latticework grill, faces a street packed with parked cars and bicycles. Inside, past a receptionist, there's a maze of gray halls leading to a line of offices. It could be a government office anywhere—say, a DMV in the States. Rooms are labeled R&D, Help Services, News Room, Network Information Department, and Human Resources.

We're meeting Bo at the new Sina office. A secretary leads us to a room at the end of one of the hallways. Inside, Bo is sitting with Zhidong, Yanchou, Wang Yan, and other Sina managers at a long table in a stifling rectangular room. The table is cluttered with smoking ashtrays, cell phones, glasses of tea, and PDAs.

The mood in the room is sober, a reflection of the mounting discontent in this branch of the new multinational. No one mentions names—probably because of the presence of a journalist—but it's clear to me that they are discussing Jim Sha's leadership of Sina and the direction he is taking the company. Just as clear: no one here approves.

Zhidong looks stressed, with dark and baggy circles under his eyes from no sleep. In addition, his normal pixyish demeanor has been replaced by a grave angst. Wearing a new short-sleeved Sina.com logo shirt, he leans on his elbows, slumping over the conference table. When he finally looks up, he says, "Worse than making a mistake is not correcting the mistake. I have to do something. I want to make certain that I have the full support of the board. Whatever happens, I will be sure that Sina is protected."

Besides checking in with Zhidong and his other friends at Sina, Bo is in town to meet with a number of entrepreneurs who could represent RFTA's first deal. Jeffrey is his note-taking shadow. Bo's meetings involve a night of careening through Beijing. The first session, at a café, is with an entrepreneur in his early forties with thinning hair, a sparse goatee, and hollow ebony eyes. He recently returned to China from Boston, where he worked as a consultant. His plan: an Internet dating service much like Sina's Club Yuan. "Chinese are even more baffled than Americans when it comes to dating," he says. "Everyone complained about arranged marriages, but meeting your perfect match is hard."

After this meeting, we drive to the Pinganli District (old Beijing), where we enter a rust-colored building with a carved, unmarked doorway. Inside there's an old man with flying wisps of white hair in a floor-length robe stirring a cauldron of indigo dye. It's hardly the place one expects to find a virtual-reality system that can replicate alternative universes, but that's what we encounter in a back room. "You can tour a building before you break ground," the system's software designer explains, "or travel to the planet Trafalmadore and meet Montana Wildhack." To see if we get the reference, he asks, "Do you know Kurt Vonnegut?"

Some entrepreneurs, like the men behind the VR system, have obvious promise. One has an e-marketing company unlike anything in China. Another has created a Web-based search engine that compares online prices.

In July, Bo and I fly to London for the first commercial flight between England and China. As long as England held claim to Hong Kong, China had forbidden the route. Now, after a series of hard-fought negotiations, Virgin Atlantic won the route. I recently interviewed Richard Branson, the Virgin founder. When I told him about Bo, he was intrigued and invited us on the flight.

It's Bo's first time in London. We have thirty-six hours. We stop at our hotel long enough for a four-hour nap. We visit Covent Garden, Hyde Park, the Tate, Westminster Abbey. Walking along the Thames past midnight, Bo tells me a story. "Two swordsmen, emperor's soldiers in old China, are on a boat that travels across Hangzhou's great West Lake. One man drops his sword into the water. Instead of jumping in after it, he sits back quietly, unworried. His friend is panicked. 'What are you doing? That's your sword! It can never be replaced!' The first soldier says, 'There is no need to rush. I can get it later.' He takes a knife and carves a mark on the boat where the sword fell. 'I'll remember its location with this mark.' The other man shakes his head at his foolish friend, since the boat is already far from the spot, and he knows that the sword is lost forever."

Bo, his hands in his pants pockets and looking down, turns to me and says, "It's a reminder that time doesn't stop for anyone. The attempt to stop the moving time is impossible." Looking up at the cathedral, he says, "It's my biggest frustration."

Bo recalls the preparations for his first trip abroad from China when he was eighteen. He was eager to leave, but extremely afraid. At the last minute before he boarded a bus for the airport, Zhijun spoke to him. "You will thrive, Bo," said Zhijun. "Do not feel the need to rush. I have every confidence in you." Bo says that his father said it with such certainty that Bo's doubts disappeared until he arrived in California.

Bo seems thrilled by London. After breakfast, we do more walking. Next we shop. Bo buys half a dozen shirts at Paul Smith. We take a limousine to Heathrow, where the Virgin inaugural flight is ready for departure. We're sent off with a Chinese dragon and a procession of Chinese dancers and music. On board, fortune-tellers and magicians entertain the passengers and a calligrapher paints messages to our families.

Branson, with hair still as long as it was in the sixties, the beard of a troubadour, and an irrepressible joviality, chats with his parents, who are in the first row in front of us. He takes a brief nap in the compartment normally reserved for pilots, and then he sits down with Bo. Branson seems eager to understand as much as he can about the

Chinese Internet revolution and tells Bo that he would eventually like to offer Virginnet, his online store, in China. When their conversation touches on soccer, both men light up. In Shanghai, Branson's staff has arranged a match between a team of Virgin employees and China's national team. It would be a certain rout except for one thing: Virgin has arranged for the two most famous Chinese soccer players in the world, Fan Zhiyi and Sun Jihai, who play for Crystal Palace, a First Division soccer club in England, to play on its team. Does Bo want to play on the Virgin team, too?

Bo is beside himself.

The plane touches down at Pudong Airport and there is a ceremony of welcome at which Branson is greeted by the mayor of Shanghai. We're bused to the Jinmao, the mirrored tower rising in Pudong like a sentinel above the other buildings in the silvery, spiky skyline. The tallest hotel in the world, it looks like Gotham City.

Bo ducks out of the evening's Virgin festivities for meetings, but he arrives early the next morning at Shanghai's Eighty Thousand People Stadium. The only other players on the field are the Chinese stars, and the three kick the ball around for half an hour. Bo can't believe he's playing soccer with two of the world's best players. When Shen Baojun arrives, he watches from the sidelines, taking photographs.

Midway through the game against China, the Virgin team, helped by the pros and Bo, is trouncing the Chinese nationals, so Bo defects. On one play he faces off with Branson himself and kicks a spectacular goal. Impressed, Sun Jihai asks, "Which club do you play for?" Shen Baojun clicks away.

Bo and I take leave of the three-day-long Virgin fete for Beijing and head directly to Sina, where the mood is sober. Last week, China.com went public. China.com is a virtually inconsequential competitor of Sina, number thirty-nine in traffic of Chinese portals. (China.com later reinvents itself as a successful systems integrator, but at the time of the IPO it offered bland, party-approved news. Neil Taylor writes in the *IT Daily*, "Everybody save its founder seems to admit that China.com's content sucks.") While Sina has tens of millions of page views and millions of registered, which implies committed, users, China.com has no registered users. Nonetheless, when China.com went public on NASDAQ, the stock more than tripled its first day out and settled back at a level well over double the offering price. China.com raised $84 million and the company is valued at $1.5 billion. Sina, not China.com, could easily have been the first Internet play from Asia. If the IPO had happened on Zhidong's schedule, not Jim Sha's . . .

The postponed IPO is only one worry. The main one, at least as far as Zhidong is concerned, is the continued erosion of the Mainland-centric operation. Sha sees things differently, of course. His experience in the Silicon Valley makes him feel justified in his view that a global company needs to be run from there, not China. Part of the reason is the assumption—shared by Zhidong—that the lion's share of Sina's revenues in the short term is going to come from the Chinese community outside of China. Nonetheless, it's a short-sighted viewpoint, Zhidong feels. Sina's strength is its Mainland base. More to the point, the nearly unfathomable potential of the Chinese market dwarfs the rest of the world's Chinese market combined. He foresees that reasonably large Chinese-language sites could grow in the United States and other parts of the world but not at the expense of the China market, where competition is heating up between Sina and the numbers 2, 3, and 4 sites—Sohu, Netease, and Yahoo China. In Zhidong's view, the winner outside of China will be the winner of the Mainland, since the main allure of a global network is the connection for Chinese everywhere with their homeland. Zhidong says, "Bill Gates tried to turn every problem into a nail so he could use a hammer to solve it. Jim Sha tries to follow that method, too." Rather than understand and focus on the subtleties in the markets, Zhidong charges that he is attempting to solve everything with a single solution—a solution geared for the United States. At least Sha has scheduled the IPO.

July comes and goes. Resentment is growing. Now comes the news that Minister Wu at the MII has made another pronouncement, this one directly effecting Sina's public offering. The government releases specific new requirements for Internet-related companies to go public in foreign markets. The regulations mean that the entire Sina business plan, dependent on an IPO, is threatened.

The feeling that he had made a terrible, even fatal mistake continues to overwhelm Zhidong. But what can he do? A battle for leadership of the company at this time could be disastrous, which ties Zhidong's hands. In addition, the Sina board, anticipating the better-late-than-never IPO, solidly backs Sha, or so Zhidong assumes.

In Beijing, when Bo and Zhidong discuss the problem, Bo encourages Zhidong to do whatever it takes to protect Sina, even if that means going up against Sha and even if it means postponing the IPO again. Zhidong is unsure, but Bo says, "The board will back you. You must act fast. You must trust your instincts."

Zhidong decides that he has to try. At the next Sina.com board meeting, held at the company's lawyer's office in Menlo Park, Zhidong expresses his mounting dissatisfaction about Sha's leadership. Sha offers to resign. Zhidong at first wonders if Jim is overloaded and tired and lacks the confidence to be CEO, but he then decides that Sha is threatening the board. "He is saying, 'If you don't agree with me on this, and if you don't give me the power I want, I will resign,'" according to Zhidong.

It's a testy climate. Dissatisfaction is rising throughout the company, particularly in China, but everyone knows that it's a bad time to make a change. Sha has set the stage for a large bridge round of financing, the government is raising its head, and the IPO is scheduled. Sha tells the board that he will only stay under the condition that he has full authority, which shocks Zhidong. What more authority could Sha have? He has already asked President Daniel Chiang to keep away from the administration of the company, and Zhidong has effectively been cut out of any decision making. Zhidong looks around the room. Rather than bow to Sha, one board member surprises Zhidong by asking him, "Are you willing to be CEO?"

Zhidong takes a deep breath. He surveys the room. He then says, "If the board trusts me, I'm willing to give it a try."

There is a vote. They go around the room. Every vote is for Zhidong. When it comes time for him to vote, Jim Sha votes for Zhidong. It's unanimous, and Sha resigns on the spot.

CHAPTER 11

BANDWIDTH DREAM

On a cool fall afternoon in 1999, Edward and Jean sit outside in the garden at our northern California home while Stephanie and our children play on the lawn. Stephanie, wearing a Pokemon T-shirt and stretch pants, has enormous hazel eyes, bright-pink cheeks, and long black hair with sharp bangs. She speaks in English, but when she hurts herself, she cries out to her mother in Chinese.

Edward has bought a home for Jean and the children in Marin County so they can be nearer to China for Edward's commute. It is a compromise on the part of Jean, who has left behind her closest friends in Texas. Over lunch she remarks poignantly that Edward's life always has been about "journey and separation," and he doesn't disagree. "Hopefully this means we'll have more time together," she says. "Stephanie misses him so much when he's not here, and the baby hardly knows him."

Then Edward tells me that he has news. When he tells it to me, I understand that things are going to get much crazier for the Tians, not easier.

AsiaInfo has continued to grow and its initial public offering, eagerly anticipated in the U.S. investment community, is scheduled for March 2000. The company remains the largest provider of networks for Internet service and content providers and telecommunications providers. Its billing software for mobile carriers and ISPs is doing reasonably well, though the integrated business acquired from Dekang was hobbling along. (That hard-won deal has turned out to be a mistake; the business seems to have topped out at 30 percent of the market, never reaching the critical mass it would have taken to be the hoped for success.) Another increasingly profitable software package is AsiaInfo Mail Center, a commercial e-mail software system similar to Microsoft's Hotmail. It is already being used by at least 10 million Chinese. Thirty percent of AsiaInfo's business is currently in software solutions while the majority is still the building of networks for carriers, including ongoing deals with China Telecom, China Unicom, and China Mobile. Overall sales are skyrocketing, and profits are growing at a rate that should impress Wall Street.

It's why I am dumbfounded to hear that Edward has left AsiaInfo. When he drops the bombshell, I look at him closely. "You *quit?*"

I'm not the only one baffled by Edward's announcement. He has stunned his friends, angered AsiaInfo's board of directors, and confounded industry analysts. It's not just that he's leaving AsiaInfo—and months before the IPO. More astonishing is that he is leaving for a start-up designed to compete against the massive state-owned monopoly, China Telecom, the 500,000-employee behemoth with $40 billion in revenues—and that the venture is being funded by the Chinese government. It's as if the U.S. government asked, say, Steve Jobs to leave Apple to create a start-up that would be a government-owned, IT-based alternative to the Post Office. And as if, out of patriotism and a sense of the challenge, he said yes.

The offer itself is a sign of how much things are changing in China. Beijing's mixed signals and its inconsistent policies regarding the

Net have gotten increasingly frenetic over the past year or so. The government has continued by turn to encourage and discourage investments in Internet technology, and it has installed and then dismantled regulations designed to monitor users and block access to certain content and communication. But after a report written by Hou Ziqiang, a renowned physicist with the Chinese Academy of Sciences, Beijing seems to have realized that it can no longer halfheartedly support the Net. Widely respected for his work as a research fellow at the Institute of Acoustics under the CAS, Hou studied the potential impact of Western IT companies—which will gain access to the Chinese market after China's admittance to the WTO—on homegrown technology. Will Chinese technology stand a chance?

Hou argued that the country would only be competitive if it launched a new initiative to construct a next-generation Internet infrastructure on par with the most sophisticated networks in the world. He also concluded that China Telecom, slow moving and rooted in old-fashioned wire telecommunications, could never pull it off. Neither could the government's smaller telcos—China Unicom, China Mobile, and Jitong—which were focused on wireless and specific Internet applications and rife with managers from the ranks of China Telecom. So Hou offered a bold prescription. The government should found a company modeled on the best Silicon Valley success stories—an independent, entrepreneurial entity charged with building a new network. Essentially, he suggested that China do what the U.S. government did when it broke up Ma Bell to promote competition. But instead of relying on the private sector to provide the competitors, Hou's plan called for the government to create the competition itself.

Beijing is often written off in the West as if it were a single, intractable, and immovable force. But Hou's boss, CAS vice chair Yan Yixun, found four influential supporters (who would become the company's board members), each from government agencies: Sheng Guangzu, deputy governor of the Ministry of Railways; Zhang Haitao, vice governor of the State Administration of Radio, Film, and Television (SARFT); Yang Xiong, general manager of Shanghai Alliance Investment; and Jiang Mianheng, who works with Yan at the CAS. Jiang also happens to be the U.S.-educated son of Chinese president Jiang Zemin and is the supervisor of the Shanghai municipal government's

Information Technology Office. With a Ph.D. in physics from Drexel University, Jiang is widely respected in China for his support, both political and financial, of China's most advanced technologies.

The CAS, under Yan and Jiang, and the city of Shanghai, under Jiang, understood the importance of a state-of-the-art Internet. The other backers had other specific interests as well. The Railways Ministry saw it as a way to secure its position in the next generation of telecommunications in China. SARFT has long sought, but failed, to enter into the profitable telecommunications business through its cable networks. With the establishment of an Internet Protocol (IP) network, competition can spill from online services into the traditional telecommunications area. "In terms of technology, there would be no hindrance to providing traditional services such as long distance calls on the IP network," Hou said, "and the fees could be much lower than the current ones," adding, "as long as the MII gives us the license."

Backed by the impressive roster of supporters, Yan set up an historic meeting in the capital with no less than the Chinese premier. It happened on February 11, 1999, when Hou gave a passionate presentation to Zhu Rongji. With China's Internet surfers doubling every six months, Hou predicted that the information flow via the Internet would surpass that of traditional audio services in two to three years. "There is no doubt that IP services will become number one in future telecommunications," Hou said. "Accordingly, the design of future telecommunications networks should target IP services." He explained that a broadband network would solve the delays of audio and video transmissions that is typical of the current online services. The quality of telephony over IP lines is poor, for example. Voice transmission is often garbled. "The problem can be solved as long as the breadth of the transmission lines is large enough," he said. In addition, Hou explained how costs could be shrunk if broadband lines were ubiquitous; he promised a monthly fee of less than a hundred yuan (twelve U.S. dollars) for surfing the Web over the new IP network, one-tenth of the current price. "There will be no time limit in that case," he said.

In his presentation, Hou acknowledged that the project would be controversial, but his passionate appeal convinced Zhu to support it. Now he had Premier Zhu along with CAS vice chair Yan, Jiang, and the other powerful government representatives all behind his idea. Within the month, the plan was approved over the objections of China Telecom's chiefs.

But who could run the start-up? Led by Hou's advice, the directors decided to look for an entrepreneur with experience in the private sector. Peng Peng, the vice minister of railways in China, who was involved in the plan's evolution, says, "The traditional route—hiring a manager from inside another government organization—would have failed."

It was a bold departure by some of the most progressive factions in some of China's most powerful ministries and agencies. To go outside the government and look for the most respected entrepreneur in the private sector to compete against the government's own monopoly seemed like an impossible idea. "Globalization, and specifically China's inevitable entry into the WTO, required nothing less than new thinking," says Peng. Yan adds, "We have seen how entrepreneurs lead revolutions and build competitive companies, so that's where we went."

Both Yan and Jiang knew Edward from their dealings with AsiaInfo, and were sufficiently impressed to put him atop their list. Credit Suisse First Boston analyst Jay Chang says Edward was an inspired, if not obvious, choice. "They looked to Tian because he has passion—for technology and for China—and charisma. He has an enormous ability to keep people focused on the task at hand. He also has a willingness to learn. But his background is not telecommunications, so his learning curve would initially be steep."

Yan asked for a meeting with Edward in March 1999. When he proposed the job to Edward, the AsiaInfo chief politely said no. "I don't think I can work for you," he said, "My company is a year away from an IPO. I have my employees to think of." When pressed, Edward was honest about his qualms about working for the government. "I'm American trained. I learned there that the mission of a company is to deliver value to our shareholders, create a strong company for our employees, and provide good product and service for our customers. I don't know how to work with a mission other than that."

After the meeting, Edward told Bo, "Every single thing about it seemed ludicrous."

Yan didn't give up. He knew how to reach Edward. Over a luncheon in Shanghai, Yan said, "Edward, your generation is lucky. You've been to the United States and you have received a good education. It's true that you used it well and you are now running a very successful company. But what about the country where you were born? We need you."

Edward argued back, "I already came back to China to help our people and I'm doing that at AsiaInfo; AsiaInfo has done good things for China."

Yan persisted. "We need models for others to learn how to run companies," he said. "Sure, you're successful at AsiaInfo, but China needs its Andy Grove, its Hewlett and Packard, its Steven P. Jobs. This is an opportunity to be that for us. As much as you can accomplish at AsiaInfo, you can accomplish more at this company—linking people's lives with broadband means unlimited exchange of knowledge."

Yan promised a free hand and a company that could mirror a Western start-up, but Edward reminded Yan of his other concerns, including the AsiaInfo IPO. "I have a responsibility to our investors and employees," he repeated. "How can I leave at a time like this?" Edward didn't stop there. "I respect your vision, Mr. Yan, but I see many problems, even if it wasn't for AsiaInfo." He said that he worried that the venture might be doomed from the start. He feared that success could actually be more dangerous than failure at this start-up, since he would be showing up China Telecom. A friend had warned him, "You'll have to worry if you fail and worry if you succeed too well, since you will make others look bad." Worse, Edward said that the proposed structure of the new company—hammered out in the high-level meetings in Beijing—seemed untenable to him. The four shareholders—the Ministry of Railways, the city of Shanghai, SARFT, and CAS—would each have a quarter interest in exchange for a $5 million investment. In addition, as a telco the company would be regulated by a fifth agency, Minister Wu's MII, which not only regulates but also runs what would be its biggest competitor, China Telecom. Translation: five demanding bosses whose interests could—maybe would inevitably—contradict and conflict with one another.

Edward also referred to the struggle of China Unicom, the government's previous attempt to create a competitor to China Telecom; the state's monopoly blocked Unicom's access to vital infrastructure. According to *Interactive Week*, "Only after the Chinese State Council intervened in 1995 was China Telecom forced to provide interconnectivity to Unicom's networks, though Unicom still had to go through

MPT [the MII's predecessor] for approvals. . . . 'Stymie your telecomrade' was the mantra at China Telecom." Meanwhile, the government prohibited Unicom from going outside for new investments. In 1998, the Chinese congress reorganized the government, slashing eleven ministries and half of the government's employees. The MPT was transformed into the MII, which was responsible for both telecoms. Minister Wu Jichuan, the leader of the new ministry, successfully argued that coordination of the two competitors would allow for more effective implementation of telecom infrastructure. Competition would be encouraged by breaking China Telecom into four companies, but real competition never materialized and Unicom still had to struggle against a competitor that controlled 85 percent of the market.

Things were changing, Yan said. The chill of the Belgrade embassy bombing gave way to pragmatism in China. Anticipating the coming WTO talks in Seattle, China and the United States put many of their differences aside and negotiated a long list of trade issues. On November 15, a breakthrough trade accord was signed. The impact on the telecom industry in China will be great, since foreign investors will be allowed to hold a 40 percent share stake in telecom providers when China enters the WTO. Two years later, the amount will grow to 50 percent. The agreement set the stage for real competition in the telecom industry, since telecoms will be able to raise foreign capital, necessary if they are going to successfully compete with China Telecom.

The experience of China Unicom was enough to make anyone wary of the Chinese telecommunications industry, though Unicom was now doing much better. The revised Chinese regulations meant that China Unicom would be able to go public in the United States. (When it did in 2000, the company was valued at nearly $30 billion. Armed with cash, it successfully carved a niche for itself in the market. It's becoming increasingly successful as a mobile carrier with a thriving and fast-growing wireless business.)

Edward continued to argue with Yan, but couldn't help but be influenced by his main arguments. Yan knew Edward enough to push the idealistic vision. Besides the technology that would allow inexpensive access, including inexpensive telecommunications (such as IP-based long distance) and bandwidth that would complete Edward's dream of building "a pipeline of free-flowing information," here was a chance to explode a monopoly that was crippling China. Edward couldn't help but be drawn into a David-and-Goliath scenario in which he was David.

On a plane flight between Beijing and Shanghai with James Ding, Edward discussed the job offer. James had to admit that the vision behind the new telecom was inspiring. "If we want to build truly world-class companies in China, we need an economic environment in China that we don't yet have," he said. "One thing we've learned: It will never happen without bandwidth. I don't know anyone who could do a better job with this, but we must understand the implications for AsiaInfo."

Edward agreed to attend a meeting with representatives of the government agencies, where he came to realize that each of the founding shareholders had something significant to offer. The Ministry of Railways could handle the physical work of building a new network as well as provide the right of way along its rail system. (Similarly, in America, billionaire Philip Anschutz purchased the undervalued Southern Pacific Railroad and later used its rights of way to lay fiber-optic lines for his Qwest Communications.) SARFT already had access to China's 80 million cable TV subscribers, an impressive initial user base by anyone's standards. The city of Shanghai's Jiang was both influential and a valuable technical adviser, with a keen understanding of the importance of information technology in the building of modern China. The Academy of Sciences employed many of China's best scientists, had a powerful and respected leader in Yan, and had a remarkable success in another attempt to spin off a for-profit company. Started under Yan with talent from the Academy of Sciences, Legend Computer did what everyone in China thought was impossible when it wrested away the leadership of the domestic PC market from formidable foreign brands such as Compaq and IBM. Continuing to gain market share, Legend had more than a 27 percent share of the Chinese PC market.

The directors of the government agencies impressed Edward, particularly when they agreed to consider his conditions that challenged assumption about Chinese bureaucracy. They included a hands-off policy and Western-style incentives for his employees.

Following the meeting, the scope of the project increasingly kept Edward up more nights. Alone in the quiet, Edward thought, Could it be done? It's impossible, isn't it? To build a new telecom for China? Me?

He kept returning to the potential of the enterprise. This new company could do more than any other single business to transform China. It spoke to his prime motivation for returning to China in the first place. AsiaInfo was integral—building the Net in China. But the new venture would create an Internet that could carry all of the information, education, access, and opportunity possible. In addition, it could ultimately be a quicker route to reach the people of China—not merely the wealthy elite. After the decision was made, Jay Chang put it this way: "If somebody came along and said to you, 'I want you to create what could become China's largest company, you'll have all the relevant government support, and you can in the process contribute immeasurably to the people of China,' what would you say?"

By then Edward was becoming more and more excited by the grand vision of an IP-based Chinese telecom, but he remained cautious, and there were a series of contentious negotiations. For one thing, there was great resistance to one idea he insisted on: stock options. As a sign of goodwill—to prove that he wasn't insisting on stock options for personal gain—he promised to give his share to charity, but options, he said, gave employees a personal stake in their work. The shareholders ultimately agreed to give 1.5 percent of the company to CNC employees.

Edward also told the investors that he would only run the company as a real company, which means that he would have to raise money beyond their seed round. The telecommunications agreements in the pre-WTO rounds of negotiations meant it was possible. However, there still was resistance to the idea that the new company had to be profitable, a novel idea for a state-owned enterprise. It wasn't that the potential investors didn't like the idea of making a solid return on their investment, but they were distrustful of the independence inherent in the idea.

That the government agreed to all of his conditions is remarkable, says Duncan Clark. "A one hundred percent state-owned company with a private-sector/U.S.-modeled management style regarding corporate governance? That is, no one shareholder can ride roughshod over management? In China, there is not a long tradition of separation of shareholders from management even in the private sector. Too often a shareholder will want to show its power to other shareholders or to management, even hobbling or destroying a company in the process."

In spite of the assurances he won, Edward still anguished over the decision for weeks. Most of his friends thought he was crazy to consider it. For one thing, they didn't trust the government to allow him a free reign. He struggled over the idea with James, who was sometimes encouraging and sometimes tried to dissuade him. "Working for the government was diametrically opposed to everything we learned as entrepreneurs," said James. "AsiaInfo is moving in a very good direction."

When they heard about it, some AsiaInfo board members were horrified. "You cannot!" said Daniel Auerbach, a board member who represented an early AsiaInfo investor, Fidelity Ventures in Hong Kong. "Edward, how can you even consider this! You are ruining your reputation! This is absolutely a crazy idea. I don't want to hear another word about it. It is professional suicide!"

Arguing on the other side, however, were his parents, who said that Edward could not turn down the opportunity to serve his country. Feng Zhijun was also in favor of the move. "This is an historic calling," said Feng. Ultimately it was this argument that carried the day. In May, Edward nervously decided to accept the position. Yan says, "Edward saw this for what it is: a chance to build technology that will help China leapfrog the West with a state-of-the-art broadband Internet backbone that is second to none. He saw the potential impact of broadband—how it more than any other technological breakthrough could change an entire country. And he saw how the WTO compelled us to build a competitive environment early so Chinese companies would be in a position to compete globally."

He and James had to discuss AsiaInfo's future. When Edward said that he wanted James to take over as CEO, James said, "Absolutely not. Absolutely not. Don't even suggest it."

"Who else would you have do it?" Edward asked. "We have good people, but anyone but you would cause morale problems. You take the leadership. If you need me, I'll be your consultant. No one else can do it but you. There's no one I trust and the people at the company trust as much as you." James said he would stay with AsiaInfo, but he felt the company would have to embark on a search for a new CEO. When he consulted with Bo, he, too, said that James should become AsiaInfo's CEO. Again, Edward said, "James, you could do this. You *should* do this."

When Edward individually told the AsiaInfo board members— and they saw that he had made up his mind—he was told to start the CEO search. "You draft a job description," said one board member. Edward said, "That's unnecessary. James should be CEO." Bill Janeway, whose Warburg Pincus was an investor in AsiaInfo and who sat on the board, said, "How could you do this to your partner? Abandon him and then saddle him with this thankless job!" Janeway threatened a lawsuit. (After Edward decided to go, Janeway relented and wrote a letter of congratulations.)

Edward said that he disagreed. "It's a good opportunity for James. He wants to grow and here he will grow. It is the best for the company. No one can lead AsiaInfo better than James. He understands this company. He *is* this company." Edward said that he felt that James had strengths that he, Edward, didn't have. "James is more logical. He is more technology savvy and he understands numbers. He's very good at strategies." Since Yadong had recently left the company to become an investor in Internet start-ups out of Shanghai, Edward continued, "We need someone who is a founder of AsiaInfo to be its leader. There's no one else with the skills, the love and respect of the AsiaInfo employees, and the passion for AsiaInfo."

James wasn't yet convinced. He said that his skill set didn't include management at the top level, but Edward said that he knew he could do it. When the board came around and agreed with Edward, James finally and reluctantly said he would try. He insisted that he spend a couple months before the transition in the United States at the UC, Berkeley, program for executives, but he had made the decision: James would be the new AsiaInfo CEO as well as its president.

Edward was ready to start his new job. However, now that he had been courted, some representatives of the government investors began to have second thoughts about him. One questioned whether he was the right man for the job, because of his youth (he was thirty-six at the time) and mostly because he was educated in the United States. It harkened back to the Cultural Revolution, when anyone educated abroad was suspected of disloyalty.

When his patriotism was questioned, Edward was infuriated, since patriotism is what drew him to consider the job in the first place. Feng Zhijun calmed him and said that the obstacles only meant that the calling was more important—besides running the new company, his role was to break down walls, to change age-old paradigms that were hurting China's development.

Finally, a deal was ironed out. The investors agreed on Edward's conditions. Autonomy. No second-guessing. Their support. The ability to raise foreign capital, including a possible IPO in the West. A share of the company for his employees. It was announced that he was leaving AsiaInfo, that James would take over as CEO, and that Edward would become the CEO of China Netcom (CNC), a new telecommunications company, a completely new type of company in China: a telecom owned by government investors that would be run like a Western-style start-up. Based entirely on IP technology, it would take on the gargantuan China Telecom.

After springing the news about his new job, Edward convinces me to visit him in Beijing, where his new office was a single room in the Friendship Hotel with a staff of two. "It's pathetic," he admits. It is. There's a stained carpet, puce-colored walls, and furniture with fake-wood veneer that is peeling off. The fax doesn't work and the printer doesn't link to the computer. When he visits, James says, "Edward! What are you doing? You left AsiaInfo for this?" To make matters worse, the government's initial funding is slow to come in and Edward asks James to borrow 2 million renminbi (RMB), the equivalent of $220,000. James says, "Don't make this a habit," but he gives the money.

Edward makes good on the loan in July when $25 million comes in from the investors. He also secures a $600 million line of low-interest credit from the Chinese Construction Bank; in exchange, CNC agrees to build the bank a state-of-the-art network.

His first chore is building a team. Hiring is a challenge, because many of the best and brightest in high tech in China have their sights on private U.S.-style start-ups or international brand names; most are highly skeptical of anything to do with the government. He lobbies a couple of candidates who turn him down, but Edward is nothing if not a good salesman. "I have become a convincer, trying to convert them to join me," he says. "I learned from studying Steve Jobs that people must feel like they're creating something. They must feel passionate, that this is an opportunity of a lifetime. And you can't convince others if you yourself aren't convinced."

His pitch is compelling, and he clearly believes it. "Join me," he says. "We can do this for the 1.3 billion people of China." Po Bronson in *The Nudist on the Late Shift* wrote that Silicon Valley "offered a chance to leave a mark on a world that already seemed terribly marked up." Even compared to Silicon Valley, China seems like a blank slate, and Edward tells employees of Western companies, "After ten years of working for Hewlett-Packard or Microsoft, have you really made a difference? Here you will have an impact. You need to do something in your life that is at least a little bit crazy. If we succeed, we will have something we can tell our children. We will leave our mark on the world." He says, "It's not only the opportunity of a lifetime, not even of a generation, but of a period of history that has the chance to create a positive future for our people."

As a legendary figure in China, Tian himself is a selling point. Says Elaine Wu, who comes aboard to work closely with Edward in investor relations, "He is one of the most important heroes. Many of us have come here because of the opportunity to work with Mr. Tian on a project that could change history." Edward promises potential employees a new type of company in China, and not merely because of the stock options and its lofty mandate. "It comes down to the way we talk to our people," he says. "This is a company in which we talk about sharing the success of our venture, always using 'we' in terms of our goals. We have open and two-way communication between everyone, no matter their level. And we are managing by building common goals and then allowing individuals to find their own path to accomplish them." An engineer who comes aboard from China Telecom tells me, "To be part of the vision of Mr. Tian, we feel as if we are on an important quest to change our country. At China Telecom, there were no rousing speeches, no feedback from our managers, no reviews or rewards, or no sense of a mission. We had jobs, plain jobs, like everyone in China. Here we have an undertaking that can change China."

Though AsiaInfo had three hundred people on staff, the company had no human resources manager. Edward's first hire at CNC is a manager of HR. "We didn't know better at AsiaInfo and so we had to build a culture after the fact," Edward says, "but your people are everything in your company. At CNC, I want to do it right the first time. We are starting with HR to bring in great people and make sure they are happy here. In weekly employee meetings we will talk about the culture. We will have extensive training. Our managers all learn what we value here—to rule by trust, not fear."

An early hire is Chareleson Zheng, a bristle-haired, square-faced engineer who was convinced to leave his GM position at Marconi, a British telecom supplier, to become Edward's Chief operating officer. "My friends thought I was crazy to work for the Chinese government," he says. "But I'm not working for the Chinese government. I'm working with Edward. Edward sold me on his bandwidth dream."

He convinces another engineer with an idealistic political story. "Broadband will bring China and the West closer," he says. "When President Clinton and Jiang Zemin have broadband-based videoconferencing in their respective offices, they can talk together in a very intimate way. If a crisis comes up, they will be able to communicate as two human beings." Says Liu Chuanzhi, the chairman and CEO of Legend Computer, "He assembled a handful of software and hardware geniuses—the most impressive minds in their fields, after many people said he would fail." Indeed, Edward successfully lures executives from Motorola, Microsoft, and Intel, as well as key engineers from China Telecom and China Unicom.

Within months, Edward moves CNC and then moves it again. The New HQ is not in Zhongguancun; it's on the ninth floor of the Copley Building, a gleaming silver structure in Corporate Square, on willow-lined Financial Street in the Xicheng District. He has a spartan office decorated with photos of Jean, Stephanie, and baby Frances in small silver frames. There's a workstation on a rosewood desk from which he can see the Beijing financial district. From the corner window, there's a view of the main street with its long line of lights with yellow globes like bunches of balloons.

Edward offers a tour of the office. There's a curved conference room where a great buzz of work is going on. We peek into the office of the HR manager, who is hiring twenty people a day. In a partially installed demo room, huge monitors show a video of workers in yellow smocks laying fiber in trenches, in subway tunnels, and along railroad tracks. In another room there's a war map like the one Edward had in his office when he launched AsiaInfo. The lines on this map represent the cities wired by CNC fiber. The lines follow the terrestrial cables that crisscross China, connecting all of the major South China cities and some far-off outposts. A dotted line indicates the coming undersea cable.

Now CNC looks similar to a lot of US start-ups, with engineers working late hours, arsenals of Nerf guns, and an espresso bar. One of those present at a midnight work session says that he hasn't received his salary because of the bureaucratic approval process, which Edward has yet to smooth out. (In another month, it is arranged.) "But you don't worry about the money when you are on a mission that is as critical as this one," says the engineer. Within two months, Edward has 250 employees. (There will be 2,000 employees by the end of 2000.)

Edward hires an advertising agency to design a logo for the company. He tells them he wants it to be accessible and green, but the one that comes back is IBM-like—mechanical blue. He angrily sends it back. "It drives me crazy," he says, explaining that green is symbolic. It goes back to Edward's days as a student in Texas, when he was planning to be an ecologist. "Green is a reminder that the Internet can help China's disastrous environment, which the industrial revolution has demolished," he says. "Bandwidth is clean." Also, the green image he has in his mind conjures up grassland, says Edward. It's another running theme in his life. "Bandwidth is grassland, a pasture in which much can grow," Edward says. "In this pasture, countless new companies and countless new applications will grow—everything from commerce companies to systems that radically improve education and healthcare."

His team immediately begins developing key strategies for the company. Its raison d'être is first and foremost to build the broadband network. With his top engineers, Edward sets about designing the network, testing and selecting the technology to use from such vendors as Lucent, Ericsson, Juniper, and others. (After extensive testing, they decide on the IP over DWDM standard.) There's much else going on, but the main effort remains the fiber build-out. "The first priority is the fiber network," says Edward. "because it is the foundation for all the products. Fiber is like a road upon which everything else will go forward." Subcontracted through Peng Peng's Ministry of Railways, a veritable army of construction workers break ground in February 2000 and dig and assemble the network, much of it along existing railway lines. It's an awesome undertaking, an almost shocking reminder that sometimes the Internet is not about brain power and virtual anything— sometimes it's about sweat. In cities, they interrupt traffic and gouge open sidewalks and roadways. In the countryside, they rip seams in every variety of field in ground that is rock hard. They send cables across rivers by attaching them to bridges or burying them in subterranean pipes, and they widen roadways and train tracks with dynamite. Thousands upon thousands of men and fewer women wear yellow CNC tunics and caps and, employing every variety of machinery and their hands, lug, chop, pour, tunnel, hammer, jackhammer, solder, break up, move and dig trenches. They excavate and pour and trowel the concrete for underground rooms for connectors and repeaters and cover them with manholes, and they lay and connect the fat black cables that come on spools the size of a man and cover everything with dirt.

Edward has teams working on immediate revenue-generating ventures, including a telephone card that allows calls over an IP network, much cheaper than calling via China Telecom or other providers' wires. (It will generate 13 million RMB [$1.5 million] in its first three months.) Other IP products are launched, including an Internet credit card, a caller-ID system for businesses, and a technology called IP-800 that connects business websites with an automatic telephone callback system. (If you click on the China Hotel website and inquire about a room, for example, there is an automatic callback from the company's reservation office to your computer's telephony system. There's no need for the less efficient back-and-forth of e-mail.)

Other teams get busy putting companies online with a myriad of high-bandwidth enterprise solutions and another group is developing CNC's datacenters. Fifty are planned, including a massive 500,000-square-foot Beijing Super Internet Center with its own power plant in Dashing County near the capital, where one hundred thousand servers will be housed. There will be a building devoted to education—showing off the broadband future to students and visitors. Servers that will fill the datacenter will carry dot-coms and be devoted to the online presence of old-economy companies. One customer that has already signed up is the Bank of China, which has attempted to manage its own technology but realizes that it cannot keep up. Edward's datacenter will power its internal and external systems and will back up everything on the bank's computer network.

When the governmental head of Dashing hears about the project, he's unwilling to help Edward through the red tape required to convert a hundred-acre cornfield into a high-tech park. Edward tries his best speech about the future of the country, but the man refuses to sign on until Edward says that the mayor of Beijing will be on hand for the signing ceremony. The man beams. "What do you need?" he asks.

Edward was right to have been worried about the complex relationship with his government investors. It takes a lot of effort to keep the officials happy. They keep their hands off, as promised, but Edward feels that it is important to keep the government shareholders apprised of his activities. "They are my bosses," he says, "and it is appropriate that they want to know what we're up to. We have designed a process to make sure we maintain the trust of the officers. There are many dinners and meetings. We have open and frank communication and we're successful at this stage."

It is worth the effort. Indeed, the government partners are partly why CNC has no real foreign competition, at least so far. CNC is also making good friends in the Chinese business community, thanks to the way it's cutting costs. "Fiber built here costs a twentieth of what it costs in the U.S.," he says. "CNC's is the cheapest bandwidth in the world." The implications of that are enormous. "With the broadband network in place comes the environment for a lot of Internet companies to very easily get started. Currently the most expensive costs for Internet companies in China is bandwidth—sixty percent of their fixed expenses. At the same time, it's six times more expensive in China to use a dial-up connection than in the U.S and income in China is more than sixty percent lower. As we get more of our network up and running and we offer more services, we can offer inexpensive access on a high-bandwidth network. As a result, the Internet economy booms. It's a snowball. More companies come in and offer everything you can imagine. More users come on, too. Everything comes from high bandwidth at a low cost. Who else can provide it?"

And yet, things may not be so easy as Edward implies. Though it's difficult to imagine successful foreign competitors hurting CNC's main IT business in the near future, CNC could face a challenge in China Telecom's nearly inexhaustible resources. If nothing else, China Telecom could make CNC's job difficult, thanks to the controls the telecom has on rights of way in some provinces, and wherever CNC's network needs to interconnect with existing China Telecom networks. On the other hand, as Duncan Clark says, "China Telecom doesn't have much goodwill with customers, to put it mildly. And I don't think the government will allow the interference; I don't think it would risk hobbling an entire industry to protect a vested interest."

China Unicom is another threat, particularly since it has a great deal of capital after its IPO. Though Unicom is focused primarily on wireless, it is entering some businesses that CNC has targeted. Another small telecom, Jitong, is also a competitor, but only in small segments of CNC's long-term business.

There are smaller competitors in China, including datacenter companies like iAdvantage, PCCW, CITIC, and a raft of niche players. And foreign competition will continue to flock in or expand their Chinese operations post WTO: Global Crossing, WorldCom, Exodus, AT&T, Quest, Deutsch Telecom, Bell Canada, Nippon Telephone & Telegraph, and Vodafone, to name a few. However, says Jay Chang, "by the time these players can enter the market with any significant strength, CNC will already have most of the important customers locked up, will have built a service organization into one that would be difficult to displace, and will have built the lowest cost network in China with upgradable conduits—something that none of the international players have access to. It's why I'm so bullish on CNC's prospects."

Fifty days after the groundbreaking, the first eight thousand kilometers in place and connected, Edward is ready to go live. The press describes it as the longest and fastest network in the world. It astounds skeptics in and out of the government. Peng Peng says, "People who were opposed to this project now see what one man with a vision can accomplish. It is waking up people throughout the Chinese government."

Edward meets with representatives of some of the shareholder ministries and then he's on the phone, checking in with the managers who oversee the building of CNC data centers, network hubs, and the laying of fiber in Gansu. From there, he takes to his e-mail, planning out a road show in the United States that he hopes will attract a significant round of private funding. In the late morning the most important thought in his head is that it's bedtime for Stephanie in California. He places a call to Jean, who tells him that the baby is asleep, and then she passes the phone to Stephanie, who has much to say about her birthday party at the Discovery Museum in Sausalito. Yes, he missed her birthday. He sees Jean and Stephanie rarely. He has seen the new baby, Frances, twice.

After afternoon meetings held in the CNC office, Edward and I jump into the backseat of a cab, which flies past acres-long Tiananmen Square, where tourists follow guides, children play tag, peddlers sell fruit ice and watermelon, and newlyweds pose for photographs. Ahead is the most famous building on the square—the Tiananmen, or Heaven's Gate, with its enormous, three-story-high portrait of Chairman Mao. Driving past it, Edward has a rare wistful moment, reflecting on the call to Jean. He admits that the distance from his family remains the toughest part of his life: tougher than building CNC at this inhuman pace, tougher than negotiating the building of the network throughout the developed and undeveloped regions of China, tougher than managing his partners and investors—tougher even than starting over from scratch. In this solemn mood, he reveals a secret, personal motivation for his ultimate decision to accept the CNC job in spite of all the reasons not to. "Broadband can do amazing things for China," he says, "and one of those things is to integrate us into the world in a seamless way. It will help to bring stories into China—stories to make dreams. Well, using videoconferencing with high-definition, large-screen monitors in Beijing and Marin, I could be in Stephanie's room in the evening when she goes to sleep. When the network is in place, I can be where I am compelled to be—here, in China—but simultaneously I can be in Stephanie's bedroom. Then, each evening, I can read to her until she falls asleep."

CHAPTER 12

MR. VENTURE AND MR. CAPITAL

In 1995, Eric Li, a youthful venture capitalist with Orchid Asia Holdings, visited Sandy Robertson in the Robertson Stephens office on the forty-fifth floor of the Bank of America tower in San Francisco's financial district. Robertson, an Orchid limited partner, told Eric, "You've got to meet my China guy," and called in Bo.

Because both were Shanghainese and both were on the China-U.S. circuit, Eric and Bo became friends, even though Eric didn't take Bo's work seriously. "I am doing twenty million dollar deals for manufacturing plants," he said. "You are doing tiny deals for miniscule start-ups with guys who had been messing around in their basements," Eric continually ribbed Bo. "Get a life! Stop wasting your time!"

Over the next two years, Bo approached Eric when he was arranging funding for SRS and AsiaInfo. In both cases, Eric declined to invest. "Are you kidding?" he said. "Those companies aren't *bankable!* Where is the value? Where is the balance sheet?"

Eric is a broad-faced man with high cheekbones. His hair, parted on the side, curves up in a wave over his forehead like the early Elvis Presley. With a picaresque smile and demeanor, he is simultaneously cautious, methodical, and fearless. He is also awesomely bright and monumentally stubborn.

Over drinks late at night in Beijing, Bo explained what was both the conundrum and the opportunity of the entrepreneurial boom as it applied to China. "In a way, you're right," Bo said. "Early stage IT companies in China aren't bankable for traditional capitalists." It's what gave Bo, as an investor focusing specifically on start-ups, an advantage. "Meanwhile, these start-ups have the potential to create enormous value out of very little capital."

Eric argued back, "The reason they aren't bankable by traditional capitalists is because they aren't sound businesses. The guys you're backing have never seen a balance sheet. They don't understand market share, barriers to entry, assets, liabilities. . . . They don't know the fundamentals of capitalism."

Bo grimaced. "Maybe you're right," he said, "but they are creating China's future, making something—something important to our country—out of nothing. Young engineers are going to have more of an impact on China than all of the shipping or steel or *mayonnaise* companies you're financing." (At Orchid, Eric financed a joint venture that brought Best Foods into China.) "Just what China needs," he said with unhidden derision. "Meanwhile, my *unbankable* guys are creating a new economy, where ideas and information—knowledge—change everything."

Eric's superciliousness began to give way. A couple things helped. When Eric was a graduate student at Stanford Business School, he became friends with Jeff Skoll, his editor on the school newspaper. Skoll went on to cofound eBay, the Internet auction house. Originally no more bankable than Bo's companies, two years after its founding eBay had a valuation of $45 billion—that's billion with a "b." It baffled Eric, but he couldn't ignore those numbers. Over the same period of time,

Eric watched SRS and AsiaInfo—the companies he had disdainfully passed on—grow into Chinese IT powerhouses, equivalents of EDS (Electronic Data Systems) and Yahoo. He wasn't giving in, but he could no longer dismiss Bo's thinking out of hand. "But how can you analyze the companies you find?" he asked Bo on another occasion. He was genuinely perplexed. "In the type of investments I do, there's an analytical methodology and you can't go wrong. You take a company, evaluate its income stream going forward based on a rational set of assumptions and projections. You discount that back to a net present value and know how to invest. With your start-ups, there's nothing but air. A guy with an idea. No analytical methodology fits. How can you decide? It's a crapshoot!"

Bo laughed at him.

"You're laughing at me!" Eric shrieked. "I can't *believe* it! You are tossing aside my reasoning, my logic! How can you do that?"

Bo laughed again.

Eric steamed. "Whether it's a good investment depends on whether the business, over a long period of time, can achieve above average return on capital," he said. "You compound that. You build barriers to entry into your business so you can thwart competition. You continually seek new structural advantages. Market share. A brand name you have earned because you have invested a lot of money in marketing and advertising. Distribution channels that other people can't crack. Economy of scale—that is, you're making ten million of this stuff and your cost is lower than the next guy's. These are structural advantages! Without those things, how can you invest?"

Bo again tried to explain. "In the knowledge economy, a young man or woman or a team of people can take a small amount of capital and their brains and over a very short period of time create a business that will compete with Fortune 500 companies and win. Why? They don't have the same structural advantage, right? They don't need it. What's the differentiating factor? The Internet. Everything is different. Everything is immediate. On the Internet, you can create a brand online without a hundred million dollar investment and sixty years. Yahoo did it. Amazon did it. eBay. How? An idea and smart people to execute it. A brand name is built overnight. A distribution network becomes less relevant. Yes, structural advantages are still relevant, but over time they become less relevant. Everything rests on what the man can do."

Eric talked to a colleague at Orchid who knew Yahoo cofounder Jerry Yang. The guy asked Yang, "What's Yahoo's advantage? Anybody can do what you do. Of course, your brand name has become an advantage, but it grew so quickly that it is conceivable that another company with a better product—something newer—could come along and create a new brand name and take over. So tell me: What's Yahoo's advantage?"

Jerry said, "Three months."

It was an instructive story for Eric. "Three months!" Eric told Bo one day after he heard it. "That's it! You can't sustain a three-month advantage with capital. You can only sustain it with people. The ingenuity of a group of people makes you or breaks you. How you structure and build the interaction among these people is what it's all about. Yahoo's constant evolution is its advantage, and the evolution is the product of its people and its culture. That's why capital is less important."

"It's not that capital is irrelevant," said Bo. "Capital is one commodity. But my dollar is only better than his is if I bring other value. Value-added capital is what it's all about. In the deals I've done, I think the founders would tell you that the money was far less important than the rest of what I could offer. I tell every potential founder that I'm not his investor, I'm his partner."

The ideas, as they continued to gel, threw Eric, who was only in his mid-twenties, into an early midlife crisis. "I'm sitting on piles of capital," he said, "but these kids are giving me a run for my money. They've got less money but can do more. I am doing something wrong."

He began to look at Bo in a new way. "I am unlearning everything I have been taught at business school," he said. "Something's happening, though I don't quite know what. I have to think in terms of advantage—why one company will make it and another won't. If it isn't a structural advantage, then what? Capital isn't it. You only need a very small amount of capital, and you can conceivably compete with somebody with ten thousand times more. Therefore, the answer always comes back to the people. The people. With the guy and his team, investors can create something out of nothing." Therefore, Eric realized, "It's behavioral advantages rather than structural advantages." He wondered, "How can I compete in a world where behavioral advantages are key? Is there a way to measure them?"

The conversations lead to an examination and reexamination of the motivation behind his work. While Bo never pretended to be uninterested in making money, it was never his predominant motivation. Bo saw that capital could do more than start businesses and potentially bring in sky-high returns. After numerous conversations over the course of the past few years, I have come to understand that Bo viewed capital as a catalyst. Inspired by his conversations with Bo and others, Eric began building on the arguments. "Five or six hundred years ago, China was the preeminent nation on earth, with the largest GDP by far," Eric told an associate at Orchid. "Nobody could compete with China; we had the highest standard of living, the strongest empire, the strongest military, unsurpassed culture. Over a period of just fifty to one hundred years, the rules changed. Why is that? China never caught up in the industrial revolution. We're still trying. However, I have seen that we can leapfrog that revolution and in so doing leapfrog the West. The Internet came when the United States had already had a very efficient business infrastructure. The Internet provides a better solution, but in the West there's a lot of vested interest in the way things were. That is, there's structural opposition to new technology. In China, however, the Internet proposes solutions where no solutions exist. There is no structural opposition. The result? Inconceivable gains."

Eric had come full circle—was infected, almost as if with a disease. His mind reeled. He *got* it.

Eric was born in 1968, which makes him Zhidong's age and a year older than Bo. Like so many other Internet entrepreneurs and financiers, his parents were teachers. It's unlikely that it's a coincidence; the children of intellectuals suffered during the Cultural Revolution, but they also were imbued with curiosity and determination, and their parents emphasized education at a time the nation de-emphasized it. Eric's father, a professor of electrical engineering at Shanghai University, had written broad-minded and respected academic papers. As a result, he was a visible target of the reeducation of intellectuals and in the late 1960s was sent to LuYu Chou, a notorious work camp in an abandoned salt mine that was once the site of a prison for murderers. Before she, too, would have been interned, Eric's mother fled China for the United States, where she enrolled at UC, Berkeley, and studied architecture.

Like so many of his peers, Eric—born Li Shimo—was raised by his grandmother. His education in grade school consisted of political brainwashing. After the Cultural Revolution ended, however, his father returned to his post at the university and his mother returned to China. Determined that Eric should be educated in the United States, she brought him to California in 1986 when he was eighteen.

He says that his name predicted his precarious position in the United States. *Shi* means "worldly." Indeed, he was being educated in the world beyond China. But *mo* means "silence." When he arrived in the United States, he couldn't speak much English.

Shimo's mother enrolled him at Berkeley High School. In her tiny second-story apartment, they sat together to choose an English name for him. Perusing a book of famous names, he stopped at Elvis.

"Who's that?"

His mother asked, "Do you know him?"

Eric didn't.

"He's a singer," she explained. "He's pretty popular."

The next entry in the book was the Norwegian navigator Eric the Red.

The "red" part of the name is what struck him. "I guess that's me," he said.

Though he didn't speak much English, Eric was infatuated by America and voraciously studied American culture. On the first day of class, he told his history teacher that he couldn't understand him. The teacher instructed Eric to come to his class after school every day for tutoring. Eric learned English by learning, with a Chinese-English dictionary at hand, the *Federalist Papers*. "I could recite them line by line but couldn't order in a restaurant," he says. After two months, he began on *Democracy in America*, by Alexis de Tocqueville. Now he tells young entrepreneurs in China to learn English the way he did. "Language is the texture and fabric of a civilization," he says. "Language connects everything. Language is the foundation. The most effective way to learn a language is to go to the source of that culture. Through the political and cultural foundation of that civilization you learn the language that is embedded in the nation. It gives you an intuitive and instinctual grasp of the words, and therefore the people."

He entered UC, Berkeley, where he studied economics. After earning a degree, he planned to return to China, but his father wrote him a letter and advised him not to "squander the opportunity" he had in America: "Get work experience. Bring that back to China."

"My life is about stumbling," Eric says, but it seems as if he is far more determined than to leave much to chance. He got a summer job in a New York investment bank. While there, he was sitting in a café reading the *Wall Street Journal*. The left-hand column story was a profile of Ross Perot. Eric was impressed by Perot's biography, particularly his rescue of EDS employees in Iran. Perot, who at the time was the third-richest man in the United States, was starting a new venture, Perot Systems. The article said that Perot had taken some of the best managers of EDS after selling his original company and was founding a "start-up." It was the first time Eric heard the term, and he decided that if Perot was "starting something up" it meant that he would need help.

Eric wrote Perot. "Do you need help?" he asked. "I have just graduated from college and I want to work for you." Next he called Perot Systems in Texas and asked for the company's fax number.

At a Kinko's off Telegraph Avenue, he filled out the forms required to send a fax and paid the $3 fee. He went home and waited for the telephone to ring. When it didn't, Eric was perplexed. "I couldn't understand why he didn't call."

The next day, he sent the fax again. He tried again the next day. And the next. And the next.

After sending seventeen faxes over the course of seventeen days, Eric's telephone rang.

"Is this Eric Li?" The voice had a Texas twang.

Eric said, "Speaking."

The man on the other end of the telephone shouted, *"Who the hell are you? You're jamming up my fax machine!"*

It was Ross Perot.

After scolding him, Perot told Eric that he might as well meet him. "Get down here to Dallas and we'll see if we can do something with you," Perot said.

Eric got on a flight the next day and took a taxi to the Perot Systems office. There were about twenty people at the firm, including Perot and the company's president, Mort Meyerson. After meetings with Perot and Meyerson, Eric was hired. His first job was as a gofer, but he learned so quickly that within six months he was in charge of the installation of an information system for a Dallas hospital.

In 1993, Meyerson, who had become Perot's CEO, offered to send Eric back to school to earn a graduate degree. Eric enrolled at Stanford Business School. Classmate Jeff Skoll was the managing editor of the school newspaper. Eric became a columnist, railing against "PC cops," the enforcers of political correctness in the business school. "There were diversity forums," Eric says. "But we quickly learned that the school wanted anything but diversity when it came to opinion. I pointed out the hypocrisy."

At Stanford, Eric met Len Baker, the legendary venture capitalist who was a partner at Sutter Hill Ventures. Baker was a renowned investor in a number of industries, including media, semiconductors and semiconductor equipment, biotechnology and medical equipment, and software. Before joining Sutter Hill, Baker worked for Cummins Engine Company, a manufacturer of diesel engines, and for the Delta Foundation, a community-owned, minority business development firm in Greenville, Mississippi. A trustee of Yale University, Baker served on the Investment Committee and on the Advisory Board of the Yale School of Management, and was a member of the Advisory Council of Stanford Business School.

Baker is an innate and effective teacher, and he regularly gathered Stanford Business School students at his house for seminars, which Eric attended. After other students left, Eric would stay behind and talk with Baker about business and art. Baker was one of the first to hear that Eric's stay at Stanford had led him in an unexpected direction. "I had planned to go back to Perot Systems," Eric says, "but I kept thinking that I should go back to China. China needed me more than America. I decided I wanted to develop a career through which I could make a contribution to my country. China was going through a growth process. I wanted to help."

Mort Meyerson expected Eric to return to Perot Systems, but he encouraged him to visit his family in China. His mother had returned, and he hadn't seen his father for a decade. After the visit, Eric was determined to change careers. Meyerson supported the decision and introduced him to Peter Joost, who offered him a job. (Edward Tian had come to see Joost when he first tried to fund AsiaInfo.) Working as a venture capitalist with Joost's Orchid Asia Holdings, Eric began shuttling back and forth between China and the States, making a series of profitable investments, primarily in manufacturing companies. "It was what China needed at the time—capital for its economy," Eric says. Orchid first invested in multinational companies that were going into China. They had the expertise and scale but didn't want to put up all the capital themselves. One was American Standard, the parent of the Trane Company, the world's largest commercial air-conditioning maker. Orchid became the largest outside shareholder of Trane China, which eventually had sales in China of more than $200 million a year. Eric represented Orchid on the Trane board. Another investment was in Best Foods International. Though Best Foods' attempt to sell mayonnaise in China bombed, other consumer products, including chicken bouillon cubes, became enormously successful.

While he worked with Orchid, Eric kept in close touch with Len Baker and kept him apprised of his slow conversion to the new economy. Eric came far enough that he began seriously thinking about going out on his own to invest in Chinese technology companies. Len encouraged him—unsolicited, he said he would invest—but strongly encouraged Eric to get a partner. "It's a lonely business," he said. "You need someone to bounce off ideas and share the workload."

In mid-1999, Eric approaches Bo. Now that Eric is convinced that Bo isn't off his rocker investing in entrepreneurial companies, he knows that no one is as well positioned or as respected. As or perhaps more significantly, the two men have grown fond of one another. "It could be crazy," Eric says, "but I think we can pull it off." Bo is startled by the offer. It's been five months since he joined forces with Sandy Robertson in RFTA.

When Eric explains the idea, he notes that he has commitments from Baker and also his friend Jeff Skoll. He feels sure that Mort Meyerson will invest, too. Bo can't help but laugh about Eric's turnaround. "So now you want to start backing companies that aren't backable . . ."

Bo is intrigued. First of all, he respects Eric. It would be an exciting, well-balanced partnership. In addition, Eric's vision for the industry has come full circle, so that it is now very similar to Bo's. The fund he's proposing is the type Bo would like to raise once RFTA becomes more established. The problem is Sandy. He has committed to Sandy, and he tells Eric that he has to say no.

Ross Perot and many others know that Eric Li doesn't hear no. He relentlessly pesters Bo with daily phone calls over the next two weeks.

Bo analyzes his position. The deals that excite him—the promising companies he has been discovering that need his money—are ones that are more appropriate to an established VC fund. Until RFTA raises its own fund, Bo has to be the banker on deals similar to when he was at Robertson Stephens. If he were to raise a fund with Eric, they would be in a position to invest directly and immediately in extremely early stage (as well as later-stage) companies. It's why Bo wonders if he should consider discussing Eric's offer with Sandy. Succumbing to what must be Eric's fiftieth harassing telephone call, Bo sets up a meeting and flies to San Francisco.

When he arrives in October, Sandy is in the office of his new company, Francisco Partners, which, with a $2.5 billion fund, is the worlds' largest technology-focused private equity fund focusing on buyouts, recapitalizations, and management-led reorganizations. It's located at One Maritime Plaza on the twenty-fifth floor, next door to the consulate of the Republic of Lithuania. Sandy moved in along with the mementos of his career—the photographs and the tombstones.

Bo begins by telling Sandy that he appreciates the support he has shown him with RFTA more than he can know. "You changed my life," he says. "You have taught me more than anyone." (It's heartfelt. More than once, Bo has told me that he owes more to Agnes Wang and Sandy than anyone other than his parents and Heidi.) Sandy listens carefully when Bo describes Eric Li's offer. When he finishes, Sandy looks at him calmly. "You're right," he says. "Bo, you should follow your instincts. If I can teach you anything, it's that. I know you have a difficult time trusting yourself, but every time you follow your instincts you're exactly right."

Before they leave for lunch at the Banker's Club, Sandy tells him that he agrees both that a pure VC fund sounds like the right approach and that Eric Li seems like an ideal partner. "I'm with you," Sandy says. "Count me in as your first investor."

They shake on it and head out to lunch. Afterward, Bo calls Eric. "Let's do this," he says. RFTA is folded after its brief life of five months, but it was an important step for Bo. Without the self-confidence brought on by Sandy's support, which led to RFTA, plus Bo's initial work building the company, he may not have been ready to be a founding partner of the brand-new VC firm.

At 12:30 in the morning, the telephone rings in Eric's hotel room in Beijing. Sandy is calling.

Eric rouses himself and asks, "How are you, Sandy?"

"Bo told me about your new company," he says. "It's the best idea I've heard in years. You guys will make a perfect team. Go and make it happen." Robertson says not to worry about RFTA. "Don't let that slow you down," he says. Sandy says that he will help line up other investors and provide references.

Later in the week, Bo and his family arrive at our house in a minivan. A *minivan!*

Bo spends the morning bouncing names for the new company. One is "BE," his and Eric's initials. Another is "Becoming," because, he says, "our job is to help in the becoming of entrepreneurs and new companies." He says, "Nietzsche celebrates 'the innocence of becoming,' because of the joy inherent in the process. We will, too."

Whatever they decide, Bo and Eric's VC fund has an impressive start. Its first investors are two of the most respected in the business. Len Baker and his colleagues at Sutter Hill decide to put up a majority stake—$6.5 million. That money and Sandy's million and a half add up to far more than $8 million. Baker and Robertson's blue-chip names are invaluable. Combined, Baker and Robertson have invested in hundreds of companies now worth much more than $100 billion, and they give Bo and Eric instant credibility. And there's more than the clout that comes from their names. Sandy and Len are hands-on mentors as well as investors. Both are motivated by their belief in Bo and Eric as much as— probably more than—the potential to earn high returns. One would be hard pressed to find more sage, knowledgeable, or hardworking advisers.

As Eric originally noted, Jeff Skoll and Mort Meyerson have already agreed to invest, too. The company's initial strategy is to start with a small fund of $50 million. They will use the money to invest in very early stage companies. The size of the investments will be decided on a case-by-case basis, but most will range from $500,000 to $1 million in exchange for between 30 and 50 percent of the companies. Smaller investments aren't worth the great amount of time it will take on each company, since Bo and Eric plan to become their entrepreneurs' partners, not just investors. Bo says, "We will only invest in the number of companies in which we have the time and energy to add value. That's our founding mission."

The partners discuss how they will carve up their duties. "Bo is Mr. Venture. I am Mr. Capital," Eric says. "That's why it was a meeting of minds, skills, and capabilities. Bo and I define venture capital. I'm good with finance, understanding how the capital markets work and how to use capital efficiently. Bo understands entrepreneurs, gets inside their skin, loves hanging out with them, staying up all night with them talking and planning. I talk to an entrepreneur for an hour or two and then, like, 'Don't you have another meeting?' Bo's like, 'Let's go for a beer!' "

It's true. Bo loves the people themselves as well as the ideas. Eric agrees to meet with an entrepreneur if he seems promising, but he doesn't want to waste his time. As Eric says, "Bo is constitutionally incapable of turning down a meeting with an entrepreneur. He's physically, biologically incapable."

Therefore Bo will search out entrepreneurs, and Eric will raise money and structure the investments. Both men will research the companies, with the help of analysts they will hire, and they will both work with the companies once the investments are made. At that stage, Bo will help companies with their vision, positioning, marketing, and management and Eric with their financials, including budgets. Both partners will help recruit talent, an exceedingly important part of the value added brought by VCs, according to Len Baker. He explains that top candidates for positions at start-ups are more likely to leave established companies if a respected venture capitalist is involved: "It's somewhat less risky than jumping onto some entrepreneur's pipe dream."

They balance one another with their styles as much as their talents. Bo's appraisal of Eric and himself sums it well: "Eric is the white gloves guy. He's the best guy to solve conflict. He's a politician. I'm the one who throws a match in the dried grass. My view is that only conflict can expedite the process of determining the leaks, the weaknesses." Eric adds, "I'm a rifle-shot guy. He's a shotgun guy." In the morning, both men may show up to work with ten things to do. Eric meditates on the list for a few minutes and chooses the three most important things based on their strategic importance. Eric tackles these three and forgets the other seven. Bo, on the other hand, looks at the list of ten things to do. "Oh, shit," he says, "there are ten things I have to do." Bo meditates on the list until he figures out a way he can do all ten of them and, while he's at it, cooks up five more.

Both Eric and Bo are relatively new to the side of their business that happens after investing—helping manage companies. However, Eric's work with Perot Systems, as well as his time on the board of the companies he invested in at Orchid, and Bo's hands-on experience with Sina and AsiaInfo should help. Len Baker says that experience is less relevant than dedication and instinct because "every problem at every start-up is unique. There are no formulaic fixes. The people who expect to find answers in textbooks will fail."

It is the first week of November 1999. The next step is to raise the rest of the capital and set up their operation on both sides of the Pacific. They head first to China to organize a temporary office in Shanghai at Orchid. Besides hiring a core staff, including Jeffrey Jiang from RFTA, they prepare a presentation for fund-raising. They have hardly touched down in Shanghai when Len Baker calls from San Francisco to say that Dave Swensen, the head of the Yale Endowment, will be in Silicon Valley in four days. Swensen has agreed to meet with Eric and Bo. "The meeting is on Friday," says Len. "I'll see you here."

Yale, the leading institutional investor in Silicon Valley, may be the most important institutional investor in America. The endowment is known for its cautious due diligence and sound management and is, therefore, a benchmark for other investors to come aboard. Swensen himself is a legend, the author (literally) of *the* book on institutional investing. (The book, *Pioneering Portfolio Management,* is a bible in the industry.) Len is anything but reassuring when he tells Eric that Yale, a lead investor in Sutter Hill, plus Sequoia Capital and many other established venture funds, rarely if ever invests in first-time funds and is even less likely to jump into a fund led by two inexperienced investors who have never before been partners. Eric thanks Len, but curses after hanging up. "SHIT! We're not ready!"

Bo leaves China on Wednesday and Eric plans to follow on Thursday. While Bo flies home, Eric puts together a rough presentation and e-mails it to Bo, who retrieves it when he lands in San Francisco. On Thursday, Bo reviews it, tweaking it and adding color, while Eric catches the last possible connection out of Shanghai that will get him to San Francisco in time for the meeting. The flight, which stops in Beijing and Tokyo, takes fifteen hours, so there's a lot of time to fret. *This is the most important institutional investor. We don't even have time to rehearse. This is going to be a disaster. We fucked up! We're fucked. We're dead.*

Late on Thursday night, Bo heads to Kinko's, where he prints and binds a half-dozen copies of the presentation. Eric is due to arrive sometime after midnight, fewer than eight hours before the meeting with Yale.

Eric calls from the plane to say that his flight is late, so he will have to taxi to Sutter Hill directly from the airport. Bo nervously drives alone down Highway 280, copies of the presentation in a briefcase.

Sutter Hill is located inside a small office complex on Page Mill Road in Palo Alto. Bo, looking serious and sharp in a dark suit, is escorted into an airy conference room where Len is sitting with Dave Swensen. Eric is nowhere to be seen, and Bo has a pit in his stomach thinking that he may have to wing the presentation on his own.

Just as Bo is about to begin, Eric bursts in. His hair is normally nicely combed and parted on the side with a bouncy wave up front, but today it's in disarray. He is pale and slightly wild-eyed.

After handshakes, Bo takes the lead in the presentation, but Eric leaps in and contradicts Bo's opening statement. Bo looks at him as if he's nuts. Next Eric takes Swensen through the bound presentation, with Bo adding his commentary. When Eric begins to relax, the presentation goes smoother, but it's unpolished and inconsistent. Swensen asks a series of questions about the fund, including the management fee (3.5 percent), the likely size of investments ($500,000 to $1 million), and the competition in China (no one is focusing specifically on early stage technology start-ups, though most of the world's major investment companies are looking at China). He also asks about the risks specific to China—regulations, political instability. Bo and Eric manage to answer Swensen as if they know what they're doing; somewhat miraculously, at least according to Eric, Swensen seems impressed by the duo. Though he is noncommittal and reiterates Len's point that Yale hardly ever invests in such untested and risky funds, Swensen tells Eric and Bo that they are onto something. He also offers some free advice: the name Becoming is cheesy.

Swensen thanks them for the presentation and says, "We'll get back to you."

The two leave the conference room and head downstairs. When they are in the Sutter Hill lobby, Eric turns to Bo and says, "We screwed up. We are *TOTAL FUCKUPS!*"

Bo is the reasonable one for a change. "We can only get better," he says. "It was our first try." It's typical Bo and Eric, who tend to alternate their discouraged moods and take turns being cheerleader. Bo says, "He says that Yale never invests in our type of fund, but he seemed interested. Maybe he won't invest this time around, but there's always the next fund."

"There won't be a next fund if we keeping fucking up like this," Eric says. "He only met with us as a favor to Len."

"It doesn't matter. It was good experience. We learned what not to do. There's a chance—"

"Cheesy?" Eric interrupts. "Our name is *cheesy!*"

"Yeah," Bo gulps. "I guess we should change it."

Before they leave the building, they decide to drop Becoming in favor of the Chinese translation of the word, which is *chengwei*. Bo says, "It's not cheesy in Chinese."

At least the decision about the name lifts Eric's spirits. They may have blown the most important presentation they will ever have to make, but they have a cool name.

"Let's go have a beer," he says.

"It's still morning," Bo argues.

Eric responds, "Not in China."

At 2 P.M. the next day, Eric's cell phone rings. Dave Swensen is at San Francisco International Airport leaving for New Haven. He says that he just hung up from his team at Yale and has arranged a meeting there. "How is next Tuesday?" he asks. "They all want to meet you. I hope you can make it."

They spend the weekend refining and practicing the presentation and then, on Monday night, board the red-eye flight to JFK, arriving in New York at dawn on Tuesday. They drive to New Haven and find that the endowment's office on the gothic campus is set in a stately room, where they present around a mahogany table.

After questions, one of the Yale team asks for references, which is a good sign. Eric and Bo know that their references, including Robertson and Baker, as well as Mort Meyerson and other respected names, are strong. There is even a discussion about the possible size of a Yale investment: $5 million.

Afterward, Eric calls Len and tells him that it went well. Len congratulates him but, managing their mounting expectations, once again warns that there is a low probability that Yale will invest. However, he says, "After Yale, your calls will be a piece of cake."

Eric and Bo return to the Bay Area and begin their courtship of more investors in the United States and Canada. They meet with Mort Meyerson and leave with a commitment of $5 million; Jeff Skoll ponies up $2 million. Len Baker sets up a meeting with BroadVision's Pehong Chen, who seems inclined to invest, although he is noncommittal. Other important institutional investors such as Stanford University, the University of Michigan, and the Power Corporation of Canada agree to meet with Bo and Eric. The partners' song and dance improves through practice. Many of the potential investors ask the same question: "Have you met with Yale?" "Is Yale in?"

After dozens of meetings, Eric and Bo see that there is a reticence about investing in Chinese technology, but there is also an even greater eagerness. An article in *Time Asia* sums up the climate, quoting Johnny Chan, partner in Techpacific.com, a VC firm out of Hong Kong. "The risks of investing in China's Internet are just this side of ridiculous," he says. "But the potential rewards are enormous."

Eric and Bo come to anticipate the questions they'll receive. Some are generic questions that would be asked of any VC, but many are specific to China: "Is this too early to safely invest in early stage Chinese companies?" "What's the government's position on foreign capital?" "Could you guys ever get kicked out?" And of course: "Could the government decide to pull the plug on the Internet?" Bo and Eric practice patience. They even respond evenly when one potential investor asks, "What happens if there's another Mao Zedong? Then what?"

It's an exhausting week—what week isn't these days?—and Eric and Bo are in Eric's Audi, speeding along Highway 101 to a meeting in the Valley when the phone sounds. Bo hears Eric's sober tone and sees the discomposure on his face. He deduces that Yale is calling with a fat no. Eric appears pained, and his side of the conversation is stilted.

"Yes."

"Hmm hmmm."

"I see."

His face reconfigures. Doom gives way to disbelief.

"Really?"

The car swerves into the next lane. Bo grabs the wheel and gives Eric his sternest glance.

"You mean we got it? We did? Are you serious? *Ten million dollars?*"

He hangs up the phone, looks at Bo. "We're going to dinner tonight."

Bo says, "Watch the fucking road."

They drive in silence. Ten million dollars is twice the figure that had been discussed in the meeting. The car sails as if it is soaring above the freeway. When they drive past Oracle's twin cylindrical towers, Bo says, "We are going to build Chinese companies that will take their place in the world like Oracle, HP, Sun. . . . That will make Chengwei the next Kleiner Perkins!"

Eric says, "I'll be Kleiner and you'll be Perkins!"

Bo looks over at him with a tight smile. "No," says Bo. "*I* want to be Kleiner and *you* can be Perkins."

"I'm afraid not," answers Eric. "*I'm* Kleiner . . ."

Following Yale, other investors seem to be pushing money Chengwei's way. Over the course of the next couple weeks, Bo and Eric travel to the coasts for back-to-back meetings with a wide variety of fund managers. After the presentations, Eric goes into "close" mode, which means pushing for signings on the dotted lines. Stanford and Michigan come in at $5 million apiece. After that, the Power Corporation invests $6 million. Along with Sandy Robertson, Mort Meyerson, and Jeff Skoll, the individuals who invest include BroadVision's Pehong Chen; Ray Chem's Bob Halpern; Jo Mei Chang, the founder and CEO of Vitria; Howard Charney, cofounder of 3Com and Cisco's senior VP; and John Glynn of Glynn Capital. Another name on the eclectic list: Donald Rumsfeld, formerly Richard Nixon's ambassador to NATO, Gerald Ford's chief of staff and secretary of defense, and Ronald Reagan's

Middle East envoy. (Rumsfeld will soon become the nation's next secretary of defense.) The fund is oversubscribed, and Bo and Eric decide to cap it at $60 million, up $10 million from the original plan.

While the paperwork is finalized, Eric and Bo return to China, where they set out to add to their initial team and search out companies (or revisit companies that Bo had discovered during his months with RFTA). Besides Jeffrey Jiang, a second analyst is hired, then another, plus office help and an office manager with impeccable credentials. In addition, a financial controller and an assistant are installed in a rented office suite in San Francisco.

There are more moments of optimism and confidence paired with ones filled with angst and terror. When Bo expresses his abject worry over the responsibility of a $60 million fund, Eric assures him that they will invest with caution. "We'll create great companies and make our investors a hundred times their money," he says. When Eric worries aloud about taking millions of his friends' dollars, it's Bo who gives the pep talk. "The money will come," he says, "but that's not what it's all about. I feel as if everything I've done is a preparation for this moment. We're finally in the position to make a real contribution. If we succeed, we will change the landscape in China."

Eric glumly interjects, "And if we fail?"

Bo answers, "I'm a good waiter."

CHAPTER 13

LET ONE HUNDRED START-UPS BLOSSOM

Shanghai in winter is relentless with a numbing and aching cold. The sky is sunless, chalk gray, when a driver meets me at Pudong Airport. He says that Bo is waiting at the Kerry Center, so we head into town. In 2002, it will be ten minutes on a train from the airport to the center of Shanghai on a high-speed rail system currently being built by a German and Chinese joint venture. It will go five hundred kilometers an hour. For now, we go about four.

When after two hours we arrive at the metal-and-glass high-rise, I look all around the marble-floored atrium. Bo isn't there, so I call his cell phone. He's finishing up a meeting in the café in the lobby.

I head over and find him sitting on a velvet couch that surrounds a Moroccan table with a chess set on top. After catching up and a couple shots of espresso, we ride the elevator to the sixteenth-floor suite of offices, which is opposite CNBC's Shanghai bureau. Chengwei is still temporarily housed in Eric's former Orchid office, a suite with polished mahogany walls and sweeping views. A half-dozen employees buzz about the reception area and more congregate in a small conference room.

Bo's corner office looks out over the former Russia-China Friendship Center with Lenin's star on top. The office is handsomely wood paneled. There are a few antiques, including Bo's desk, which is covered with a week's worth of newspapers. A large monitor is connected to a tiny Viao, which itself is connected by a dial-up to the Net. (Chengwei's new office, in a building being constructed a few blocks away, will be connected to a T1 line.) On a shelf behind Bo's desk are a dozen pictures of Heidi and Tiger. On the floor, there's Tiger's car seat, which is poignantly empty.

Bo joins the meeting in the conference room, listening in on presentations by his analysts about the progress of their research on a few companies he and Eric have identified. Chengwei has already made several investments, mostly companies that Bo found while he was at RFTA. One investment is a relatively small $300,000 in a company called Textclick, which is an online book publisher. Several other investments are imminent. The ones that sound particularly promising to me are a community website called Renren, an online security company called InfoSec, and Red Flag, a leading Linux company founded by the former head of Microsoft China. Beyond those, the range of companies under consideration is wide, from Web-hosting businesses to online translation services to wireless software companies. Chengwei also makes an investment in AsiaInfo, which is arranging a final round of funding before it goes public; James Ding offers Bo and Eric the opportunity to invest. It's not the type of investing Chengwei plans, but it's a safe opportunity and they put in $500,000.

After the meeting, we jump into the elevator. "We've been in business a few months and we have a strong portfolio," Bo says. "We have to be careful not to overinvest, so we can give each company the attention it needs. The hard part is saying no." Bo says that there are many good ideas and strong entrepreneurs. "The challenge is to find the ones that will deliver," he says. "Delivering means a type of discipline that many entrepreneurs lack."

Outside the Kerry Center, Bo throws on a heavy coat over his suit and leaps into the car, which groans as it moves into the frozen traffic. He shakes off the cold with a shiver and speed-dials Eric, who is in his car in San Jose. No matter where they are, though they often are on different hemispheres separated by at least thirteen time zones, Bo and Eric talk at least twice and sometimes as many as a dozen times a day. They use e-mail for letters, documents, and quickly typed notes, but the telephone is their lifeline.

After small talk, Bo stops short. "You're *what?*" He holds the phone away from his head and tells me that Eric just hit someone.

Bo asks, "Is anyone hurt?"

After being reassured, he tells Eric to call him back if he needs anything.

Bo hangs up. "It's a fender bender. Eric is heartbroken. I thought he was going to cry. It was his brand-new Audi."

The phone rings again. It's Eric again. He's all right. He says that the driver of the other car was on his cell phone, too.

Eric once said that Bo is constitutionally incapable of turning down a meeting with an entrepreneur, and I see what he means when Mr. Venture hits the streets. Bo is on the lookout for companies who deserve Chengwei's hard-won cash. Bo arrives at a session with the heads of InfoSec, a company that is creating Web-based security systems, including innovative encryption technology. The potential security market is enormous in China, and there is little competition. In addition, foreign competition post WTO is likely to be hampered by the government; trade agreements or no, the government won't make it easy for VeriSign, an analogous U.S. company, or other foreign security companies in such a politically sensitive sector. On the other hand, InfoSec has its unique challenges. While creating secure systems for companies and consumers, it will have to work with the state security agencies. The unanswered question: While it's clear that the government will allow technology that protects users from hackers and fraud, what about its own prying eyes?

Most of the research on InfoSec is complete, Bo says, and he and Eric are negotiating their investment. The founders are incredibly talented and highly skilled techies who don't have much business experience, which makes it a complicated negotiation. Originally the founders wanted 10 million RMB (about $1.2 million) for 40 percent of the company, but Bo and Eric calculated that the price was too high and the equity too little. Their current offer: 6 million RMB (about $700,000) in exchange for 50 percent of the company. The founders have thus far resisted the deal, so Bo has to play tough with the amiable men.

"It's Chengwei's final offer," he says, couching it as gently as he can. "It's the only investment that works for us. We want to work with you, but of course you don't have to take it. Anyone who looks at the risk will be at least this cautious, however. You have enormous promise. However, you have a long way to go before you'll become profitable. A lot can happen during that time."

Before Bo leaves, they reluctantly agree to the deal.

While Bo and Eric's respective responsibilities are loosely defined—Eric crunches numbers, Bo assesses the caliber of the people and the general strength of the business plans—both partners do just about everything that needs to be done. However, they both must sign off on every deal. Many of the greatest success stories in Silicon Valley may not have made sense on paper. Venture capitalists invested because of their gut. However, Eric and Bo know that the problem with going on instinct is that Bo's is somewhat overeager. It's not unusual for him to become excited about an entrepreneur and want to sign a deal before he looks at the books. That's when it's Eric's job to calm him down. That said, they know that there are times when the books don't reveal the full picture, and Eric slowly becomes better at trusting Bo when he concludes an argument, "We should gamble on this guy."

Back in the car, Bo explains the process that he and Eric go through before signing any checks. "We've developed rigorous stages," he says. "Part of the process is designed to evaluate the business, but the underlying investigation is into the people. We want to see if they have the ability to think through things rationally. To determine the level of commitment. Sometimes you can't see that at the first meeting or the second. But you may see it at two in the morning when they have their guard down."

Bo continues, "We explore their business plans with them, even though we know that the plans may change in two months. It's not about whether this is a plan we're going to stick with for three years. Our name implies that it is about what they will become. However, a sound plan proves that they have the ability to think about the issues correctly."

Bo says that he was forthcoming with the InfoSec team. "They could become VeriSign," he says, "but they could disappear without a trace. High risk, high reward. It's what makes it insane. It's why we're here."

Bo next has a meeting with someone from Renren, a company Jeffrey Jiang is researching. The company is one of China's most popular online communities, similar to Geocities in the United States. (Geocities was acquired by Yahoo for $4.7 billion in 1999.) Bo was at RFTA when he met the founders of Renren. When Bo learned the name, he was practically ready to write a check on the spot. *Ren* means "person." *Renren* means "every person." After months of an intensive investigation, Chengwei becomes part of a group of investors. They put in a million of a $5 million round. (It will turn out to be a mistake. Renren will crash along with the dot-com bust that is on the horizon.)

China is behind the United States in terms of payment systems and other factors key to generating online revenue, so Eric and Bo are extremely reluctant to invest in dot-coms, particularly business-to-consumer sites. It's not that they think that portals and e-commerce won't succeed. Rather, they know that it will probably take longer for them to reach heights relative to their American counterparts. It's why the two make a strategic decision to spend most of their capital on what Eric calls "the arms dealers of the dot-com war"—that is, companies that provide e-business solutions, including Web hosting and marketing to other companies, including traditional businesses and "the dot-com warriors." The remainder will go to dot-coms such as the online book publisher and a comparison-shopping site called e135. Chengwei turns down numerous business plans that depend on e-commerce. Eric and Bo's caution will wind up protecting them from the bust that is awaiting the Internet industry.

They coinvest with Goldman Sachs in a company called Rebound, an Internet company devoted to helping bricks-and-mortar companies sell their excess inventory online. Chengwei is the minor investor, contributing half a million dollars compared to Goldman

Sachs's $3 million. They also invest a million dollars in a start-up called T2 Technologies, which creates innovative e-marketing systems. The company is already breaking even because of a software package that serves up online advertisements on more than seventeen thousand websites.

In a small second-story office above a shop that sells carved wooden Buddhas and jade icons, Bo meets a young man with unkempt hair and a shiny gray suit who has founded a Web community somewhat like Renren and, in the United States, the Well, Stewart Brand's Berkeley-based community, one of the first on the Internet. This one has a loyal user base, and the founder is looking for money to build the site.

"I know the site and I'm impressed by it," says Bo, "but there are too many Web communities." He discloses that he is an investor in Renren and says that the entrepreneur's site, in spite of its committed user base, has no revenue stream. If it tries to copy the established communities, it will be a latecomer playing catch-up.

The founder seems dejected, but he and Bo keep talking. In passing, Bo asks what else the man has been up to. He tells Bo that he made a "cool" piece of software: a portal for PDAs such as Palm Pilots. "In the West, PDAs offer clunky versions of the Web," says the man. "They just pare down existing portals. This one is designed for the small screen just with the key features that are the most popular. You can use it to learn English, download MP3 music files, check stocks, do mail, and buy stuff."

Bo asks for a look. The man shows it off on his laptop. It's a simple and elegant interface. Bo's interest level is rising. "This," he says, "is your future. Forget the community."

The man removes his thick glasses, wipes his eyes, and sits up straighter. Bo says, "This program is your company. Forget the other one. We need to do some research about the field to check on the competition, but we might invest in this."

Before the two men say good-bye, Bo says, "The first order of business is a haircut, and the suit has to go."

The man looks down at his suit and asks, "What's wrong with it?"

The man leaves and Bo is still firing. "We want people like that," he says. "We want artists. The problem is that artists are moody. If you want to succeed, do the work. This guy's got promise! My father always said to me, 'If you're gold you're always going to shine.' I am looking for the gold. This guy is gold."

Bo meets with the founders of other start-ups that represent Chengwei's first investments. One, a former manager at AsiaInfo, runs a business called SeeWAP (WAP stands for "wireless application protocol"), a wireless application software company. Bo sees it as significant that AsiaInfo, one of the first Chinese IT start-ups, has spawned a new company. It proves Edward's idea that the Chinese IT industry can grow exponentially after the first wave of companies are established. There are two reasons. Entrepreneurial companies train future entrepreneurs, which is particularly significant in a country with no entrepreneurial tradition. Second, successful entrepreneurs are a new type of role model in China. Before now, young Chinese watched corrupt party cadres succeed. There were Taiwanese and Taiwanese Americans Barry Lam, Charles Wang, Jerry Yang, and Pehong Chen, but who from the mainland? Now they look to the likes of Edward, James, and Zhidong.

Bo also meets with an earnest man, solidly built, who has curious eyes. His name is John Sunn, and he is the founder of a company called Oval. Sunn, who is thirty years old, studied electrical engineering at Dartmouth before working in finance at Merrill Lynch and J. P. Morgan. Sunn left to found, with four friends, a dot-com called Ordermyfood.com. Wall Street workers could peruse menus of, and place orders with, local restaurants on the website. The restaurants paid a fee to be included. When the business needed cash to grow, John went to California to look for funding. He met up with a friend named Yangdong Shao, a Stanford student who spent a five-year stint at Solomon Brothers and two years working at Intel. John says he was "railroaded" by Yangdong when he proposed an alternate venture.

Yangdong was an eager and whip-smart business student who, studying the IT business in China, came up with a powerful idea. For post-WTO China to be competitive, its industry must be as efficient as industry in other parts of the world. However, most Chinese companies, whether state-owned enterprises, spin-offs, or joint ventures, have never been effectively managed. Many still are run on paper, while others have archaic computer systems. Western companies, including Oracle, BroadVision, and PeopleSoft, could automate them. However, there are major obstacles for their products and services in China. For one thing, their software is too expensive for many companies. For another, it's designed for Western companies that usually have centralized operations. Most large Chinese companies are highly fragmented with multichannel distribution systems and large numbers of suppliers, with many moms and pops on both sides. How can such haphazardly organized companies become more efficient? That's where Yangdong's solution comes in. Web-based systems could pull together a company and its thousands of partners. Anyone with a PC and an Internet browser could be connected. That's why it's possible to imagine the entire manufacturing base of China leapfrogging from the dark ages to the most advanced, versatile, completely scalable, completely upgradable, and affordable system in the world. John and his team had the talent to create it. If they pulled it off, they could create a company that would dwarf Ordermyfood.com.

They worked together on a business plan at a Chinese restaurant in Palo Alto and consummated a deal at the Stanford Oval, which is where they found their name. They designed their own logo, and were excited about it until they realized it looked just like the Gatorade logo. Back to the drawing board.

Even before Oval opened its doors, the technical whizzes on the team began to design a program. Next Eric, who knew Yangdong, set up a meet with Bo. The Chengwei partners were so impressed with Oval and Yangdong Shao that they made an unusual offer. They agreed to back the company, but they wanted Yangdong to come work with *them*. Negotiations with the various parties took three months, but a deal was cut. Chengwei invested $2 million in exchange for a 35 percent share of Oval. For John and his team, Chengwei would provide more than the operating cash. Oval needed help setting up and operating in China. In the meantime, Yangdong became Bo and Eric's junior partner.

Soon after the deal, John and his original team plus four hired engineers set up in a Shanghai warehouse in a rummy part of town. The cheap rent attracted them. There was one large room with a fresh coat of white paint. It was empty except for tables with workstations connected to a DSL box and a watercooler. The team worked feverishly writing software.

When we meet at Oval, John tells Bo that he has been pitching

Oval's software to several of China's largest companies, and there have been some bites. First Auto, the largest company in China, and Goodbaby, the world's largest manufacturer of baby products, are interested. After working in New York City for his entire adult life, John expresses surprise at the business climate in China. He says that the head of the IT department of the second-largest television company in China told him, "If I use your system, it will kill my people's income. They rely on kickbacks from our current vendors. It will make some of my team redundant." John was shown the door. Through a contact of Chengwei's, however, John returned to the company and met its CEO. At that level, the Oval solution was greeted as a money saver, bringing more efficient work and the need for a leaner staff. No deal is yet signed, but John seems encouraged. He says that he is still making three or four calls a day while his team of engineers works eighteen-hour days. When we leave the warehouse, Bo tells me that John and his founding team pledged not to take any salary out of the business until they have their first paying customer.

I'm staying at Bo's latest home in Shanghai, an apartment in a dull orange building with sentries posted at the entrance. There is a health club and pool across a circular driveway. His two-bedroom apartment on the sixth floor is simple but tastefully furnished (a white sectional sofa, small antique tables, and a paper Noguchi floor lamp). The place feels like no more than a pit stop, which is exactly how Bo seems to use the apartment, at least until Heidi arrives. Indeed, at the end of the day, Bo stops in to change clothes. The TV is on, but he's not watching. Music is playing, but he's not listening. He checks the refrigerator and closes it. The only things inside are a couple of beers and bottles of mineral water. He checks it and closes it again as if something new might have materialized.

Bo emerges from his bedroom spry, dressed in a light-brown suit and knit shirt. He has on a pair of soft leather Gucci shoes, but doesn't bother to tie the laces. It's dark outside, but he's wearing a new pair of bronze-rimmed sunglasses. Rather than clip on shaded lenses for outside, these have magnetized lenses that seal on with a click. After a day and then another of meetings with entrepreneurs, jet lag—a permanent condition for Bo and most of his colleagues—has wiped me out. I pass out on the metal-frame bed in the extra room. Bo, however, heads off to Baojun's house to watch a soccer match between China and Korea on television. If the home team loses, it will be its twenty-second loss in a row.

In the morning, as we head out for breakfast, Bo tells me that China lost. So did Bo, who played six hours worth of chess, mah-jongg, and cards. Baojun apparently won every game. "Baojun-chen"—Bo adds the affectionate Japanese suffix *chen* for his friend—"hadn't slept for thirty-six hours," says Bo. "He was at his intellectual peak. I couldn't compete."

Bo and I are walking along a congested street in old Shanghai. Some Shanghainese wear pajamas underneath puffy jackets as they shop for fish and produce. People sit in their doorways chatting while inside their TVs are on; a group of boys, crunching *youtiao* (fried dough), listen to a CD from a boom box; and a man in plastic sandals pours a bucket of steaming dirty water onto the street.

We stop for breakfast at a streetside dumpling house. The owner, a good-natured man in a blue smock, teases Bo for his chic wardrobe, which stands out in the greasy room with rows of stained tables. The other customers are wearing grimy shirts and baggy work pants.

Behind the counter, an assembly line of cooks place teaspoon-sized balls of ground and seasoned pork onto the center of squares of fresh dough. They fold the dough around the pork and twist the ends shut, making bite-sized packages. Another cook plucks them up with long chopsticks and places them into bamboo steaming baskets.

We settle on facing benches over steaming bowls of wonton soup, freshly made cabbage buns, and glasses of soybean milk. When Bo answers his cell phone (it's Eric), the other patrons, along with the cooks, stare and listen in. They have odd, uncomprehending expressions. Bo is speaking Shanghainese, but he is throwing around figures in the multimillions of dollars.

To attempt to repair the damage done by the all-nighter, we head to the gym in Bo's office building, where a dozen businessmen and women are working out on StairMasters and treadmills. We swim, and then we sit downstairs in the lobby coffee shop and drink green tea until Bo's new driver calls us. When we meet him outside, I see Bo's new car, an elegant Audi A6 2.8. The personalized license plate is 7698, Tiger's birthday. When I examine the car, silvery like a rainbow trout, Bo protests that it is a business necessity and not extravagant. (He pays the costs himself.) He notes that it's not a limousine. He wouldn't be caught dead owning one. He explains that it is a compromise. He lusted after the impractical TT, Audi's sleek two-seater.

The driver is Xu Bing, a high school classmate of Bo and Shen Baojun's. Until this month, Xu had been driving a cab in Shanghai. Bo quadrupled his salary and promised to teach him about business. I won't be surprised to see Xu in a business suit in a year, but now he wears colors coordinated with the car's chocolate-and-cream interior: a T-shirt, brown jacket, khaki slacks, and sandals.

Baojun, now part of Bo's personal staff, is riding shotgun. As his personal assistant, Baojun negotiated the complexities of buying this car, for example, and signed a lease on Bo and Heidi's apartment. Bo still makes his own appointments and answers his own telephone, but Baojun's salary is easily justified in the time saved dealing with the many headaches of running a business in Shanghai, where it's never easy to buy a copy machine or get reliable telephone service. Baojun is a hard and resourceful worker, but the main reason Bo employs him is because he can. Baojun hasn't been working, and Bo tenderly remembers his friend's help when Baojun had a hefty bankroll from his job with the Tokyo Yakuza.

Xu drives from Shanghai onto the highway and heads toward Hangzhou, where Bo has a meeting at the office of T2 Technologies. En route, Bo carries on two conversations at once in English and Chinese. With his Chinese friend, he is arguing about the Korea-China soccer match. He's telling me that Eric will probably one day run the World Bank. "He wants to do it and he could."

"What about you?" I ask. "What will you do?"

He smiles and says, "The Formula One team." Then he adds, "But not until I have created at least one great company as significant in China as Sun, Cisco, or Motorola is in the U.S."

On the outskirts of Hangzhou we pass by characterless apartments and dreary shops and factories and brick fences that are falling down. "Sometimes I wish the clock could be turned way back," Bo says. "This was a city of history—one of the seven capitals of ancient China—and unsurpassed beauty, but Heidegger says that we are inherently out of step because we cannot stop time. We are isolated, alone, and unable to feel things as we go forward." An Air France Concorde crashed over the weekend and, reflecting on it, Bo says, "When there's a crash, we're surprised. It's more surprising that things hold up as well as they do. Still, progress is the natural way things evolve. Globalization is next, whether we like it or not. We are never stable in our lives. We are like a table with four uneven legs. We continually adjust them, which means that every other leg must change. Ship, steam engine, internal combustion engine, Internet, arms race, cloning, damming the Three Gorges—but nature will break at some point. It can't take any more beatings." A minute later, he says, "What will happen? Kierkegaard says, 'Life can only be understood backward, but it must be lived forward.' We'll look back and tally our mistakes."

Downtown Hangzhou is indistinguishable from any large Chinese city, but everything changes on the lake, where the roadways are lined by lush and dark green vegetation and buildings with traditional architecture. We check into our hotel, the Zhejiang. There are pictures in the lobby of Mao Zedong from when he stayed here, too. (Bo apparently shares Chairman Mao's taste when it comes to hotels.) We are shown to rooms that overlook what was once the private garden of a wealthy tea merchant. Then we meet in a pagoda with green lacquered "wave" tiles with eyes on the round ends. Bo, Baojun, Xu, and I sip glasses of watermelon juice.

The last time Bo and I were in Hangzhou it was for the meetings that lead to the AsiaInfo-Dekang deal. Now he has a quick meeting with the founder of T2. Afterward, I accompany Bo to play golf. Bo says, "Golf is the sport Buddha would have played." When I look dubious, Bo explains, "Golf is about the moment. It is a method for changing one's sense of time, which makes us think with an alternate perspective. It can be very useful."

In Hangzhou, Bo, Baojun, Xu, and I arrive at a country club, though only Bo and I are playing. The best part is watching Xu and Baojun, who have rented a golf cart. Baojun doesn't have a driver's license but is behind the wheel. At one point, the cart gets air over a small bridge and lands inches from the lake.

We play in the hazy dusk amid sparse palms and pines. I nearly bean some *wu* birds, long-necked and with scary eyes. There is no wind. The golf cart winds past a small tea farm with signs on posts that say not to enter the tea field because of "the snake." Then the quietude is shattered with what sounds like booming thunder. Bo explains that it's mortar fire. We are near a military base and a target practice is under way. Bo says, "They're practicing for Taiwan."

After golf, we drink *yu qian* tea. *Yu qian* means "before the rain"; it is fresh green tea harvested in spring. Then we dine at the Hongjun Restaurant, a cavernous wood-paneled room with a line of private rooms on both edges. The front room has tanks of eel, green-spot fish, lobster, and varieties of clams. We sit at an octagonal table with a lazy Susan that is soon crowded with pork knuckles, pickled yams, salted duck, boiled shrimp, and sautéed bok choy served with dishes of spiced oil. Noodle soup comes next. Then a whole fish. Jet lag hits again and I feel as if I could easily pass out on the table, but I manage to make it through more courses—little bird lemon soup with fresh mint leaves, yellow melon cooked with silver ginger, pickled garlic, lightly fired eggplant. Next: the famous West Lake, or *Xihu* soup, served with cold beans, lentils, and pickled onion.

After dinner, we walk under gold *yangliu* trees that rain over the inky lake's edge. Baojun, whose hair sticks up in a cowlick and wears size small shirts and boy-sized pants that are too large, walks with his hands in his pockets, his head down, listening to Bo talk about America. Baojun wants to come to San Francisco to learn English. Bo agrees to help him.

Everyone else crashes at the hotel, but Bo sits down in the lobby, where he furiously dials the Bay Area, checking in with Eric about progress on various deals. Maybe it is because Bo winds up so intensely that it is impossible for him to unwind and take advantage of the chance for some sleep. Maybe it's something else. In the hotel lobby, sitting in bamboo chairs near an antique bed called a *kang*, Bo tries to reach Heidi on the phone, but she's out. He becomes a little gloomy. Before jumping back on the phone, he sips a beer and says that he cannot stand downtime because he misses Tiger and Heidi so much. "What will happen," he asks, "in two years, when Tiger begins kindergarten? He won't be able to easily be dragged back and forth between the U.S. and China." Bo admits that he represses thinking about it. Heidi has said that she will move to China, but he worries if she will be happy here away from her family and friends. Will she adapt to life in China?

Inevitably, the conversation leads Bo back to philosophy and he is quoting Freud, Einstein, Buddha, and his favorite, Heidegger, again. He says, "Goldfish in the pond cannot comprehend water. We're not advanced enough to think and too advanced to live." It's Bo's interpretation of the rat race of humanity—and the futility. "We haven't transcended life with an ability to truly understand, but we no longer can simply live as animals," he says. "That's why we're always tormented." Bo says that he sometimes thinks about alternative lives. "Maybe I should become a monk."

I tell him that it may be difficult to choose between the life of a monk and a Formula One driver.

Bo shakes off the dark mood, explaining it away as jet lag. "There's the moment at 3 or 4 A.M. when you are in hell, longing to be wherever you aren't, certain that the people you love are dead, terrified of the night, anxious for the dawn," he says. At times like this you can watch Bo drift away and almost see him thinking in the vacancy of his eyes. When he's like that, he lets his guard down, but it doesn't last.

The next morning is clear and the sky is pink. We have tea in a Qing Dynasty teahouse called Autumn Moon on Calm Lake Pavilion, which is set high atop thick crimson pillars. Sitting under a canopy of red leafed trees that grow underneath ornate eaves painted red-brown, we look out over a thick bamboo forest with the lake in the distance and crunch yam sticks and eat olives with longjing tea. Whenever our cups are low, the waiter pours in more hot water in the traditional manner from several feet above, making a long waterfall.

Afterward, we get into the car to return to Shanghai. Bo is behind the wheel. He turns on the sound system and cranks up a Bartók CD. Before he drops me off at the apartment in Shanghai, he calls my wife in California. She's not in, so he leaves a message on her machine, asking if she has heard from me. He says he's kind of worried. The last he saw me, he says, was when I had gone off with a hostess in Hangzhou.

Bo and I are back in his office, which is deserted. He and Eric are conferring. Eric and Bo discuss whether they are investing too much too quickly or too little too slowly. "The U.S. Internet market has set the stage with ridiculous valuations, ridiculous IPOs, ridiculous mergers," he says. "All a company needs is a dot-com after its name and it is a player, but there will be a day of reckoning. Recklessness will not go unpunished."

Eric will turn out to be more right than anyone can guess. However, without a crystal ball, Chengwei is in a fairly extraordinary position. On one hand, caution is prudent in any business venture. On the other, there is somewhat of a rush to sign good companies in order to take advantage of the madness of the market. In spite of Eric's warning (and in spite of some early signs of trouble), the Internet boom has yet to reach its pinnacle—as the AsiaInfo IPO nears.

CHAPTER 14

THE TAO OF THE DOW

The press over the "reorganization" at Sina—a polite way of describing Jim Sha's departure and Zhidong's return to his position of CEO—is untimely given the fact that the IPO is finally scheduled. *BusinessWeek* terms the changes "a management purge." Since Sina is in its pre-IPO quiet period, the company can't respond. The press aside, there is a rekindled feeling of purpose inside the company. Sina once again has a single vision and a single direction—Zhidong's. In practical terms, it means that the Chinese market will take priority over other Sina markets elsewhere in Asia and the United States. It doesn't mean that the other markets are unimportant, but the emphasis on the Mainland helps focus the entire company. Even the managers in

Sunnyvale, including the original Sinanet founders, seem newly inspired.

Now that the high-level maneuvering is in the past, Zhidong has his work cut out for him. He's back in the middle of every challenge he faced at the time of the merger, plus the damage done to the divided company must be repaired. Zhidong has to win over or remove Sha loyalists and scrap or move forward on projects initiated under Sha, depending on Zhidong's take on them. To accomplish all this, he works tirelessly, flying back and forth even more often between the Sina offices in the United States and China, doing whatever it takes to unite the company. And if that's not enough of a task, even as Sina adapts to its org-chart tumult and recommits to users in China, Wu Jichuan of the Ministry of Information Industries reappears. His action this time, which seems aimed specifically at Sina, is particularly badly timed, and the IPO seems caught in the balance.

The Internet gold rush in the United States hasn't been lost on the ministry heads in China, who accurately gauge the great anticipation of the Sina offering. The shorthand in the investment community for Sina is "it's the Yahoo of China," even though the actual Chinese Yahoo—that is, Yahoo China—has been launched and taken its place as the number four portal after Sina, Sohu, and Netease. The association with Yahoo and the vast number of potential users in China are enough to hype Sina's promise, even in the face of the company's lack of profitability. (These days Internet companies aren't expected to make money, at least in the short term. In fact, Amazon.com is the best example of a company that has thrived in spite of gallons of red ink.) There's every reason to believe that the Sina offering, even if tempered by the appearance of chaos inside the company, will be snapped up. In fact, there are projections of an opening share price in the high twenties or early thirties.

In the MII's ranks, some worry that foreign investments could threaten the government's ability to control or even influence Sina as well as other Chinese Internet companies. In their view, it's particularly dangerous, because Internet companies offering news, commentary, e-mail, chats, and newsgroups are increasingly becoming media and communications companies. The CCP has successfully controlled media and communications, but the Internet is eroding that control. What would happen if an Internet company was controlled by foreign investors and a foreign board? It would be tantamount to a U.S. company owning the controlling interest in a Chinese television station or newspaper, a scenario that the Chinese government would no doubt prevent at all costs. It explains Minister Wu's newest edict, which is a reiteration and toughening of existing regulations that expressly forbid foreign investment in the Internet. The bottom line, as it is announced in headlines throughout the world: Chinese Internet companies can have *no* foreign investors.

Panic sweeps through Sina and its underwriters, though there's hardly time to respond. Three days after issuing the statement, Wu offers a "clarification" that sounds more like a complete reversal. "On the whole we support foreign investment in the sector," he says. There's a collective sigh of relief, though some damage has been done. If nothing else, Wu reminded investors who may have forgotten that the Internet business in China is unsettled and volatile and therefore extremely high risk.

I hear several theories that seek to explain Wu's about face. One is that he is testing the waters once again, pushing to see how much control is possible. Another is that he is sending a warning flag. The more likely behind-the-scenes scenario is that business interests in China successfully pressured Wu to make his immediate reversal, an indication that the Net is already too important to the Chinese economy to be hindered by anyone, even by the nation's minister of information.

There are more rumblings from Wu and other government representatives over the course of the next month. If not a ban on foreign investment in Internet companies, there will be limits, Wu says. The probable reason: There's pressure both to allow the Sina IPO and to retain control, particularly since this public offering will set a precedent—a dangerous one, according to some factions at the MII. Sina's chief competitors, Netease and Sohu, have also filed for IPOs. If the government can't control Sina's IPO, it will set the stage for a free-for-all.

At least the Sina site continues to thrive. On September 30, two weeks after the board meeting in which Zhidong shuffled the Sina deck, an earthquake measuring 7.6 on the Richter scale hits Taiwan, killing 2,295 people and causing billions of dollars in damage. Minutes after the tremor hits, people in every country of the world boot up Sina.com. Traffic surges to new records. Once again, the company's servers are stress-tested as thousands check in to get news—for some, news of family in Taiwan. There are more users signing up all the time, and the site is becoming more robust with the offer of free e-mail (similar to Yahoo's and Microsoft's offerings) and free home pages.

All along, Zhidong and his staff, lawyers, and supporters are working to push the IPO forward. Zhidong, riding a veritable yo-yo of plane flights between Beijing and Sunnyvale, returns to the Mainland at the beginning of October. Arriving in the late afternoon on a Friday, he races to Sina, where he meets Wang Yan, Yan Yanchou, and other executives to plan their strategy for the IPO.

On the following Monday, Zhidong and the others visit the Ministry of Information Industries, where they meet with Wu's underlings. The ministry representatives refuse to acknowledge that the latest decree is a shift in policy. They are reiterating the existing policy, they claim. The Sina team argues that the existing policy is inconsistent and confused. Why has nothing been done about the enormous amount of foreign investment in almost every Internet company in existence, portals included? What about SRS's early investment rounds, all from foreign companies (including Bo's Robertson Stephens round)?

The government responds that SRS was a software company, not an Internet company. Anyway, the past is irrelevant. The new set of regulations is meant to correct the situation and establish clear guidelines for the future. The bureaucrats insist that rather than hinder Sina, they want to assist one of China's most promising IT companies. They will work with Sina to find a solution. However, the Sina bosses must understand that there is "extreme sensitivity" about foreign involvement in this industry. Rather than helping their cause, Zhidong discovers that the WTO negotiations, which are perpetually heating up, are an impediment. The reason: The government wants firmly established and enforced regulations as a clear starting point for the negotiations. Thus far, the WTO negotiations have paved the way for foreign investments in Internet companies to peak at a 49 percent stake. It's a way for foreign companies to participate in the building of the Chinese Internet and share in the growth, but no outsider would have the ability to set policy or control a company.

In the discussions with Sina, the government raises a series of questions. "Why do you need an IPO?" asks one negotiator. "Why not private funding?" Zhidong educates his audience about the Internet environment. "To be a player, we don't just need the cash to scale, but the ability to do deals," he says. "Valuations are extremely important for that. The Internet economy forces ramping things up quickly." It's a crash course in Internet economics. Yahoo, like Sina, is a strong company and a market leader. It has kept its position not only by hard work and a superior product, but by its ability to make strategic alliances and acquisitions. Yahoo's performance on the stock market has meant that it has been able to do whatever deals its leaders have deemed important. Yahoo hasn't had to spend its cash reserves, either; as a result, it is (that was the theory anyway) protected if there is a market downturn. Another benefit of a publicly traded company is the ability to offer its employees stock options in place of high salaries. In short, Zhidong explains, the Internet economy at this early stage is dependent on the stock market. To compete, Sina needs to play.

Another MII bureaucrat asks, "Why not wait until the domestic capital market is cleaned up? You can take Sina public in China." A Chinese saying is evoked: "The best things should be kept within the family." Zhidong says, "Waiting will jeopardize Sina's market leadership, particularly since Yahoo China will have no such restrictions. The Chinese division is a branch of the foreign organization and can take advantage of its parent's strengths."

In the midst of the tense negotiations, someone on the Sina team, who has become frustrated by the process, throws up his hands and barks, "Why Sina? Why are you picking on us?" A government representative snaps back another Chinese saying. "To eliminate gangsters the most important thing is to get the leader of gang," he says. It's more evidence that the government feels it can control the other Internet companies if it can control Sina.

There's more sparring and posturing on both sides, though most of the meetings with Wu's negotiators are friendly. Throughout, the government maintains that it wants to help Sina—to support it because Sina is a good company, good for China. On the other hand, it is clearly established that under no uncertain terms will Sina be allowed to go forward unless it cooperates with the government.

The back-and-forth leads to an agreement by the government that the IPO can go forward as long as China is protected. How? The MII proposes various structures that would limit the company's activities and/or limit foreign capital, but the Sina team explains that the requirements would make an IPO moot, because Sina would no longer be an attractive company to the Western capital market.

Zhidong and the others make counterproposals even as the date of the IPO nears. "The train is racing down the track and an immediate solution is necessary if the IPO is to go forward," Zhidong says. He's met by another setback: The government announces that it wants to embark on a research project in order to better understand the Internet economy.

Zhidong is ready to tear out his hair. "A research project! That will take months!"

The IPO is postponed. Negotiations push past Thanksgiving, Christmas, New Year's, the Chinese New Year. Sina's lawyers as well as outsiders continue to try to help break the impasse. Zhidong calls Feng Zhijun to help. Bo's father and other leaders lobby for Sina as Zhidong continues to host numerous delegations from the MII. He gives them grand tours, explains the company repeatedly with PowerPoint and live demonstrations. He and his top executives make numerous speeches at the MII's offices. They provide detailed reports and explain their far-ranging vision of why Sina's success on the foreign market will not only be better for the company, but for the whole of Chinese industry. "Sina's success will inspire young men in China to start their own companies and to push them to study high-tech development," says Zhidong. "In addition, an overseas IPO means that we are using foreign money to help the overall development of China's IT industry." It's important to encourage this trend, he says, because the Chinese economy is continually strapped for cash.

In February, Sina agrees to a proposal by the MII, though there are more discussions about the details. The agreement is hardly ideal. In fact, it confuses almost everyone, including potential investors. Sina will be divided into two companies. The company on the Mainland will be a separate entity, a Chinese corporation. Sina's U.S., Hong Kong, and Taiwanese businesses will be independent. However, it's a smoke-and-mirrors solution. Sina China will buy virtually all of its technology, as well as its business services, from Sina.com, which will be a listed company in the West. Sina.com will be the sole technology and service provider for Sina China. It sounds a lot like everything on paper, but little in reality will change. Nonetheless, the deal is viewed as a triumph for both sides. Whereas it makes investors (and some staffers inside China) wary, it breaks the stalemate. The new structure loosely complies with the government regulations—the government can exert its control on the local company, which is exclusively responsible for content on the Mainland site—and Sina's past investors are protected. The IPO can finally go forward.

Across Zhongguancun at AsiaInfo, James Ding, the freshly appointed CEO, is himself preparing for an IPO. Edward's exit and James's ascent have gone smoothly. When the company's staff heard that James was the new CEO, some worried that he lacked Edward's charisma. However, as one of them says, "We were shocked at the transformation. His dynamism had been hidden, directed like a laser at his technical team. When he broadened the output to the whole company, we felt supercharged." James spent several months in the executive training program at Berkeley and came back with inspiration, largely confident that he could pull it off. He applied his software genius and obsession with making things work to AsiaInfo itself.

James is so nice, precise, and reserved that it is difficult to imagine him at the helm. Handsome and with undiminished enthusiasm, he exudes a restrained confidence and yet has lost none of his diffidence. Indeed, he is intense and without any pretension. Says Yan Qu, a local manager, "James has built a strong management team with high-level recruits from IBM, Lucent, and HP. There is enormous potential for growth, since China is rapidly wiring on every level and there isn't much domestic competition." When I ask if AsiaInfo is worried about the influx of competitors that will come with China's entry into the WTO, Minhua Shao, the company's technical director—he was employee ten—says, "Oh, no! The WTO is good for AsiaInfo, since we are the most likely local partner of a wide variety of foreign companies if they enter China. AsiaInfo has state-of-the-art technology, an excellent track record, good relationships with the government, and products and services that will be essential." Minhua adds, "AsiaInfo could easily grow ten times in five years."

The new AsiaInfo office is on the outskirts of Zhongguancun. The doors and windows are frosted and etched with parallel designs that evoke Frank Lloyd Wright. The oak furniture also has Wright's symmetry and sturdiness. There are framed black-and-white photographs by Mario De Biasi on the walls. The floorwide rooms are filled with orange-and-beige cubes. The headquarters have the air of calm, professionalism, and success.

In December, before the IPO, James sits at his desk and writes a memo to a list of AsiaInfo's "friends," including some investors. Of course there are no guarantees, he says, but things look extremely promising: "For those who don't know, Asia Resources Holdings Ltd. is an over-the-counter stock that has 'ASIA' as their ticker symbol. AsiaInfo is also applying for the symbol 'ASIA'. . . . According to NASDAQ, AsiaInfo will be able to get 'ASIA' symbol and the other company will be forced to change its symbol, because that over-the-counter stock has lower priority. However, when Reuters and Dow Jones reported our S-1 filing, they mistakenly coded the news under 'ASIA,' which was later on carried by Yahoo and other ICPs and financial media. As a result, Asia Resources Holdings Ltd.'s stock jumped 500 percent on first day of the news release and continued to surge as high as 1700 percent until yesterday when our follow-on clarification started to reach investors. AsiaInfo's counsel and PR firm issued a Press Release the second day after the filing and NASDAQ was informed and consulted about this issue. NASDAQ confirmed that our filing was correct and the follow-on approach was appropriate. But we are still monitoring closely any development of this issue. Enjoy your holidays and have a happy New Millennium!—James."

Later, before James leaves Beijing for the road show, he has dinner with Edward at a Beijing noodle shop. Edward has a moment of "a sinking heart" when he realizes that AsiaInfo is really James's now; that James, not he, is taking their company forward to this hugely significant milestone.

As the Asia Resources mistake predicted, when he opens the company's books in the historic road show (it's the first time a private Chinese company is going public in the West), investors seem eager to participate. When trepidation is expressed, it's not about AsiaInfo, but about Asia in general or China in particular, though the on-track WTO negotiations help assuage some of the fears.

James is on the floor of the NASDAQ exchange in New York on the morning of March 3, 2000, the day the stock is to begin trading. Edward is in Shanghai. There is enormous excitement for the principals in AsiaInfo. The initial offering price is $24 a share. Bo, who is meeting with Eric and Yangdong at the time, keeps in touch by phone throughout the morning. When our friend at Morgan Stanley Dean Witter calls to report that the stock is opening in the nineties, Bo is, for once, speechless. The stock quickly hits one hundred. Bo does some calculating in his head. The stock price means that Edward and James, with ownership of 18 and 17 percent of AsiaInfo respectively, are, for this brief moment, worth about $500 million apiece. The deal values AsiaInfo at more than $5 billion. It is the first private Chinese company to reach $1 billion, never mind $5 billion.

That rainy night, Bo drives to the northern California coastal town of Inverness for a celebration. His brother Tao is in Canada and plans to join us, but he isn't able to get into the country because his visa has expired. Eric plans to come, too, but he is home with a fever. So it is just Bo and me at Manka's, a lodge in the wooded hills of Inverness, where Louie, the enormous lodge Labrador, is sprawled out on the couch in the waiting lounge. I tell Bo that I am buying dinner, so he chooses the best Bordeaux on the wine list, and with it we toast Edward and James. We also toast China, which today has become a participant in the technology gold rush. Bo is feeling elated. "Today in this, the year of the dragon, China created its first private billion-dollar company," he says.

After the close of the market, when it is morning in China, James calls Edward. They congratulate one another. It's impossible for many people to believe that that money isn't important to them, but they don't even touch on their personal wealth in the conversation. They reflect on the step they've just taken, which is important not only for AsiaInfo, but for all of China. The company they built is strong and established, richly capitalized, and positioned to become one of the dominant IT players in Asia. Technology stocks on the market will soon sink, taking AsiaInfo's stock price down a quarter more, but the accomplishment cannot be diminished.

In the days and months following the IPO, the personal money becomes trickier for Edward and James. Even though their newfound riches are only on paper, the two men (like the other Chinese IT millionaires who follow) find that money is a double-edged sword in China. A million yuan, which is $120,000, is an enormous, unreachable fortune to most people in China, where the annual per capita income is much less than $1,000. A million dollars, never mind a hundred million, is inconceivable. Neither man changes his lifestyle in spite of the newfound wealth, and they downplay it to reporters. I actually get the feeling that it is embarrassing for them both and particularly problematic for Edward, who is working closely with government bureaucrats who, though twice his age, are earning standard government salaries. A year after the IPO, *Forbes* prints a widely circulated list of the richest entrepreneurs of China, which includes Edward and James in the fourteenth and sixteenth positions. Because of the lower stock price (the stock market is down and the technology sector is down most of all), each has "only" a couple hundred million dollars. Edward is openly criticized by some of those around him, which indicates an odd if predictable contradiction in China as the country races to embrace capitalism. Jealousies are logical, but it would seem that there would also be an acknowledgment of the larger issue when the Chinese look at another *Forbes* list. Li Ka-shing, a Hong Kong real estate developer worth $11 billion, is one of the world's top 20 wealthiest people. However, there is no one from the Mainland on the list—no one on the list of the world's richest 250 people in fact. Surely if China aspires to the top of the world, it needs its share of success stories.

The principals aren't the only ones who profit from the AsiaInfo IPO. The investors who were brought in by Bo's original deal do, too. Warburg Pincus and China Vest hold 10 and 4 million shares respectively after their original investments of a few million dollars apiece. For their several hundred thousand dollar investments, Edward's friend Carol Rafferty and Sandy Robertson have also done well. Chengwei's AsiaInfo shares have shot up, too.

For Edward and James, however, the successful IPO is more than anything a vindication of their efforts and a green light about the future. For Edward in particular, it's a brief moment to stop and accept that his dream is coming true. A Chinese technology company he and James created was accepted—no, celebrated—in the world market. It's the strongest sign yet that Edward is succeeding in his plan to connect China to the information superhighway. A latecomer, China has become a participant in the world Internet boom.

Just when it tanks.

Of course, we don't know it at the time, but AsiaInfo's is one of the last skyrocketing IPOs of any company anywhere, at least during that period. A month afterward, on April 14, 2000, the U.S. stock markets are hit in Wall Street's most volatile day to date. Falling technology shares send NASDAQ and the Dow Jones Industrials plummeting. And it's just the beginning of the descent. In a matter of months, NASDAQ hits its lowest level in a year and the Dow charts look like the EKG of a fibrillating heart. Amid the sinking tech stocks, Internet companies are decimated. NASDAQ soon drops to its lowest point in two years. The technology industry, particularly the Internet, spirals ever downward. The result? Fear, trepidation, lost fortunes, dried-up capital, panic. Just when China was about to become a participant and take advantage of the euphoria, it's gone. As a result, the Internet in China never gets the boom that provided rocket fuel to countless start-ups in the United States. True, many of the Western companies fail when the balloon bursts, but the U.S. Net itself benefited from the enormous rush of capital and innovation over the course of a few years. The ascent in China will be slower, but it may be more sustainable.

An immediate result of the downturn is that many planned IPOs are postponed. Companies that go forward do so at much greater risk than during the euphoric late 1990s. There are some successful offerings, but increasingly few companies come close to AsiaInfo's successful IPO. In fact, a growing number of companies fall underwater, which means that their stocks dip below the offering price.

The timing couldn't be worse, or more bitterly ironic, for Sina. Zhidong's company had the chance to beat both China.com and AsiaInfo into the public market. If it had, it would undoubtedly have raised enormous caches of capital and would have enjoyed, at least for awhile, one of the enormous Internet-era valuations. However, because the IPO was twice postponed, the offering is scheduled in the midst of the tenuous climate of the stock market descent.

Some of his advisers counsel Zhidong to wait, but the hard-won deal with the government makes it an untenable proposition. The window of opportunity feels somewhat precarious, so Zhidong, Wang Yan, Daniel Mao, and several other Sina executives set out on the road. Like James Ding, Zhidong has to calm worries about the volatility of the Asian economy in general and China's in particular. Unlike James, he has to do so in a pessimistic market when everything Sina has going for it—technology, the Internet, and China—are viewed as deficits. To make matters worse, Zhidong has to address the sticky issue of Internet content and the somewhat murky, as well as unprecedented, deal with the government. Even worse than that, he has to explain the departure of Jim Sha and other executives and the last-minute switch of underwriter. (Zhidong has brought in Morgan Stanley to take over the deal.)

It all adds a sense of the insane to the road trip, but after everything they've been through, Zhidong and his team have no choice but see it through and do the best they can. There are no expectations of an AsiaInfo-type IPO, and Zhidong's partners at Morgan Stanley are floating opening share prices that are much lower than Zhidong wants. Still, the company's record as the number one portal in what could become the world's largest Internet market remains compelling. Even in this grim climate, investors seem interested.

After the West Coast leg of the road show, Zhidong and the others drive with Bo, Heidi, and Tiger to the Cuvaison Winery in Napa Valley. The sight of a flapping bright-red People's Republic of China flag overhead meets them at the winery's entryway. After sampling Cuvaison's best, the group drives up to the Auberge de Soleil for dinner. Zhidong and Bo recall the last time they were here, when Bo tried to dissuade him from creating Sina.

There's not much time to relax, since Zhidong and his team have to head east for the IPO. In more meetings with investors, Zhidong continues to find that he is able to answer their understandable concerns about the restructuring. Moreover, they seem to grasp Sina's long-term potential, and they're impressed by such statistics as the site's rapidly growing user base, which includes eighteen thousand new registrations a day. Also, Sina's events continue to draw enormous crowds. In addition to its popular coverage of the latest Olympics, records are set when the site airs a live birth—of a baby named Dian Dian—and chats with celebrities such as filmmaker Zhang Yimou. Advertising is inching up and the Sina Mall is beginning to generate revenues.

In New York on the eve before Sina will begin trading, it seems likely that Zhidong has pulled it off. The IPO is oversubscribed at the price of $16 a share, even though this is less than half the price he once hoped for. At night, he and his team are filled with tense optimism, though they try not to think about it. Instead, Zhidong, Daniel Mao, and Wang Yan attend a New York Knicks game. Then, in the morning, accompanied by their Morgan Stanley Dean Witter bankers, they head onto the trading floor of the stock exchange. Trading begins at ten, and the stock briefly pops up and hovers slightly above the opening price. It then creeps up as high as thirty before settling above $20 a share, a 27 percent gain. Relief. No 400 percent gain, but more than respectable by any measure other than the one of the recent stock market frenzy. Even when the stock declines, Sina is valued at $900 million, and Zhidong's 6.5 percent is worth almost $60 million. He and other Sina executives are wealthy beyond anything they could ever have reasonably expected. As with Edward and James, the wealth seems meaningless to Zhidong. Nothing changes in his life other than Liu Bing becomes pregnant with twins and she retires from her office manager job. For Zhidong, however, there is more work than ever, since he now has a global public company to run in addition to the separate Chinese company based in Beijing.

The Internet sector continues to get hammered. NASDAQ dips and dips lower. Occasional (increasingly rare) encouraging earnings reports shoot a stock upward, but in the near term the descent seems irreversible. Zhidong has a nearly impossible task: justify the IPO in an environment that is devastating established, profitable companies and thrashing the new economy and unprofitable companies that were, until recently, Wall Street darlings. He's not alone, of course. Tech companies from Microsoft to Nokia to Cisco crash. AsiaInfo is sent down into the teens. The stocks of Yahoo, Amazon, and other companies that defined the Internet revolution are free-falling. Yahoo dive-bombs from a high of $250 to $13 a share.

On my first visit to China, when the Internet economy continued to defy gravity, I was told that *da kang*, which sounds a little like dotcom, means "wondrous prosperity." Bo now informs me that a slightly different character set with a similar pronunciation, *da keng*, means "bottomless pit." What a difference a vowel makes. Indeed, *da keng* sets the tone of Bo and Eric's work as they go forward. That is, what very recently seemed all about prosperity now seemed to be about managing a bottomless pit. As James Surowiecki writes in the *New Yorker*, "Gone are the days when a pair of flat-front khakis and some high-flown rhetoric about reinventing the gelato industry could get you a hundred million dollars in venture capital and a cover in the trades. Venture capitalists have become misers. Institutional investors have become cowards. And hordes of dot-commers have become unemployed."

It refocuses Bo and Eric's job. They always planned to work as partners with their entrepreneurs, but now they have to become overseers of their investments, offering gentle reminders or tough sanctions—whatever it takes—to encourage companies to refine their business plans, work harder, and tighten belts. Bo says, "There is lots of blood on the Street. There will be more before it's over. We want to be sure it isn't ours."

The market continues to sink throughout June. In July, the number two and three Chinese portals go public. Sohu.com and Neatease both come out in the low teens and quickly sink below the offering prices. The market continues to sink to the point that they lose most of their values, and there is talk that they might be delisted. Sina itself struggles in the single digits, and there is speculation of a consolidation among two or more of those companies.

There's one way that the tighter climate benefits Chengwei. When the Net-and-China combination was smoking, investors had begun to flood into Beijing flashing cash. Now they are retrenching and in some cases fleeing. Chengwei is one of the few companies committed to Chinese entrepreneurs. It means there is less competition. It also means that the wildly inflated expectations of entrepreneurs, who had visions of riches and unrealistic valuations of their companies, are tempering. Since they're more desperate for Chengwei's money, Bo and Eric can fund companies at much more realistic valuations.

No one knows if the troubled environment is temporary or not. Eric's early apprehension about the Internet economy is vindicated. "I thought it didn't make sense and it didn't," he says. "Now the old formulas are working again. We can help companies because we know what's expected."

For his part, Bo, though he's all for belt tightening and business plans grounded in solid economics, isn't quite ready to jump on the bandwagon and sound the death knell of the Internet boom. He says, "The euphoria inflated things to an unrealistic level, so there is a naturally and healthy pullback. However, the stock market doesn't have anything to do with the fundamental power of the Internet. Internet companies will make money, because the technology will become integral to the economy. There will be a hundred million people in China who will be online in a few years. Within a few years more, there will be several hundred million." He knows that it won't happen without setbacks—financial, political, and personal. And, in fact, there's another sobering episode this summer.

Huang Qi, who describes himself as "an Internet writer," lives in the southwestern city of Chengdu. Huang has been praised in the *People's Daily* and other official news organizations for his website, which helps track down missing people. The site has been responsible for uniting at least two hundred families. In addition, collaborating with the Sichuan and Chengdu police, Huang's website helped find seven teenage girls who had been kidnapped and sold for wives. They were rescued and reunited with their families. On another occasion, he was instrumental in the rescue of a three-year-old girl who had been kidnapped and sold in the underground baby market.

In December 1999, Huang launched a second website, which he called Scream Online, reachable at the Internet address 6-4tianwang.com. The "6-4" is the date of the massacre at Tiananmen Square. The site is devoted to exposing human rights violations and corruption in China. On it, Huang has posted information about the underground democracy movement, the independence movement in the northwestern Muslim region of Xinjiang, unfair treatment of laborers in Chengdu, and the repression of members of Falun Gong. In the weeks leading up to the eleventh anniversary of Tiananmen Square in 2000, Huang posted incendiary articles that called for the government to reevaluate its position on the 1989 demonstrations and to "rehabilitate" the students who took part. He wrote that the government should acknowledge that the students were patriotic heroes. Huang also published a heartbreaking and inspiring letter from a mother whose son was killed during the demonstrations.

The police arrive at his home on the evening of June 3, 2000, on the eve of the Tiananmen anniversary. It is the first publicly reported arrest related to the Internet since numerous Falun Gong arrests and the arrest of Lin Hai. Before he is dragged from his home, Huang hurriedly posts some final words on the website. "We have a long way before us," he writes. "I appreciate all of you, all those who help China's democratic progress. The police have come, good-bye!" His wife, arrested with him, is subsequently released, but Huang is not. Charged with "subverting state power," Huang, if convicted, could face life in prison.

After the arrest, a message posted by a Chinese student on another Chinese website reads, "Mr Huang's prosecution is a sign of how the Chinese government is determined not to allow the rapid development of the Internet to undermine its control on information or its monopoly on power. But our friend will not be stopped. We shall continue to use the Internet to overthrow the obsolete regime."

There is no further word about Huang, but surprising news comes from Beijing. It is first published online in a report from the Bureau of Democracy, Human Rights, and Labor in the U.S. State Department. Lin Hai has been released from prison six months early.

Richard Long of *Dacankao*, who confirms that Lin is out of jail, maintains that "my Internet friend" was persecuted because the government couldn't get to Long himself (or others who work on the dissident newsletter and website from the United States). Long says that the Internet campaign helped influence Lin's early release. Human Rights Watch, in a report published online, agrees that online publicity and protest, particularly since it was timed with China's push for admittance in the World Trade Organization and its campaign to win the 2008 Olympics, helped Lin's case. (Beijing will go on to win the Olympics in a controversial decision announced in July 2001.)

Lin Hai returns home to his family and remains in seclusion until he goes back to work designing a new website that posts job openings for Chinese firms.

In early 2001, Richard Long helps me to reach Lin Hai by e-mail. Lin says that he has left China and is living in New York City. Answering one of my questions about his background, Lin recalls when he was a student in Beijing during the Tiananmen Square massacre. On June 4, he recalls, "in the afternoon I saw an eight-year-old boy dead. His mother was crying on our campus. His father said the boy was shot last night in the street."

Lin doesn't know why he was released early, but he says he was a "model" prisoner. After his release, two policemen tracked his activities, including when he took a trip through remote China. When the opportunity arose, Lin left for the United States. "I want a chance of the life and the air of freedom," he says. He is unsure what he'll do. In New York City, he's looking for a job. In his e-mail message, Lin says that he remains optimistic about the influence of the Internet in China. "The Internet is where freedom grows," he says. "When commercial companies do business in China, they bring something else into China. . . . In China, the Internet is the only media that cannot be controlled one hundred percent by the government."

CHAPTER 15

外贸商场

ROUSING THE SLEEPING DRAGON

*When I was young I, too, had many dreams. Most of them
I later forgot, but I see nothing in this to regret. For
although recalling the past may bring happiness, at times
it cannot but bring loneliness, and what is the point of
clinging in spirit to lonely bygone days? However, my
trouble is that I cannot forget completely.*
—LU XUN, PREFACE FROM *Call to Arms*

Whee possible, Bo and Eric have their weekly DFU meeting—that
is, their Don't Fuck Up meeting—in person. However, many of the ses-

sions are, like today's, on the telephone. It's July 2000, and Bo and I are in the backseat of a car driving through Shanghai. As I've come to expect, the late summer is scorching with steam rising from asphalt streets along which road workers and gardeners toil with headbands but no shirts. We're in Pudong under silver towers that reflect the shimmering blue sky, stuck in traffic.

"Burn rate, burn rate, burn rate," Bo is yelling into the phone. "We have to control it. Burn rate will kill us."

I deduce that burn rate is the theme of today's DFU.

The reason is obvious. Some of Chengwei's initial investments are burning through cash at a terrifying pace. Renren is spending $1 million a month. Renren had a successful backdoor IPO in Hong Kong, so there was an infusion of cash, but at this rate the money will be gone in a matter of months. Compare that to Oval, which is spending only $100,000 a month. Sure, Chengwei could pour in more of its fund (in exchange for more equity) into troubled companies, but Bo and Eric have made it plain that that's not an option.

Eric plans to scrutinize the budgets of every company they've invested in. Next he and Bo will consult with their CEOs to identify cuts. Hopefully there will be a meeting of the minds, but Bo and Eric know that there is the probability of disagreements.

This DFU is over quickly. The way not to fuck up is crystal clear. Chengwei's companies need to simultaneously tighten their budgets and increase revenue. Cutting burn rate and signing up customers.

At week's end, Eric flies in from California, and he and Bo set out to meet with their entrepreneurs. They are used to playing off one another, filling whatever role is required for the specific situation. Now Bo is charming and inspiring, Eric is hard-nosed. Now Bo is the irascible bully, Eric the reasonable and compassionate voice. Both have been known to lose patience with an entrepreneur and walk out of the room, leaving the other to soothe wounded egos. "All right," Bo will say. "Let's figure out what needs to be done." The good cop/bad cop routine worked, but there's no longer room for a good cop. Sitting opposite their entrepreneurs, Eric and Bo say, "The capital we have given you is all you've got. We aren't going to give you more money, and no one else out there will write you a bigger check. You had better make it on this capital or you're gone. Given that, let's make a plan."

They pore over budgets and send their CEOs out to get customers. In some cases, Eric adds, "It's hardball time. If you are unable to do what's necessary, we will remove you as CEO. We have no choice."

Adds Bo, "It's our responsibility to our investors. We know that you're brilliant, making great technology. We also know that you are uninterested, unable, or unwilling to get customers. If you resist going out into the marketplace, however, you won't survive. Your great technological achievements will be for nothing."

"Get customers and don't come back until you have them," says Eric. "Capital does not give birth to businesses. Technology does not give birth to businesses. Entrepreneurs do not give birth to businesses. *Customers* give birth to businesses. Paying customers."

"Next week, when we come back, we want a list of the ten customers you visited this week and the ten you will visit next week," says Bo.

Leaving the meeting, Eric winds down. "Get us your budget," he says. "And go get customers."

Some entrepreneurs understand, but others resist, arguing that all they need to do is raise more money. "If you do, you'll be toast," Eric tells them. "You don't need to be wasting time dressing yourselves up to woo investors and fund-raising. You need paying customers."

After one of the meetings, Bo seems particularly drained. The fact that Chengwei has decided to exercise restraint, reining in Bo's impulse to invest in every entrepreneur he encounters, is somewhat traumatic for him. Now everyone at Chengwei, including Bo, Eric, and Yangdong, plus Jeffrey and the other analyst, is doing almost nothing but working with the Chengwei investments on budgets and business models. In some cases, they are working with their entrepreneurs to change the very premise of their company. Bo's body shows the strain of the new regimen. His shoulders hunch, his tall frame limps, and he sits back in the car with his eyes tightly shut. It's a different type of work than he signed on for. It's grueling. It's tedious. It's not exciting. At least not today. And it's impossible to know if this strategy will work. Whereas six months ago it was possible to imagine that half of Chengwei's first dozen companies could go public or become acquired within a year or two, now there is no sure sense of exits. All that is known is that in China, just as in the United States, there will be a shakeout.

Bo works throughout the day and then we meet his brother, who recently moved to Shanghai for his new job. Tao is in a long-sleeved carmine shirt. He has crew-cut hair and a handsome, strong face and thoughtful eyes. Tao and Bo look dissimilar except for their eyes. Both contain an unwavering dark luminance.

Tao has followed Bo to become a venture capitalist focusing on information technology. In contrast to Chengwei, Tao's company, New Margin, is backed by the government of Shanghai. It's another venture initiated and backed by the influential and forward-thinking Chinese Academy of Sciences' Yan Yixun and Jiang Mianheng, two of the men behind CNC. Tao seems revitalized. Physically he is unchanged except for his haircut, but the big difference is in the speed he moves and the look in his eyes. Undeterred by the weak stock market, he has been infected.

As enormous as the Cultural Revolution was in shaping Bo, it had a greater impact on Tao. Born in 1967, Tao's early memories include humiliations by the Red Guard. His mother was too ill to care for him, so he was sent to the "rehabilitation through labor" camp with Zhijun. He worked digging rocks out of hard dirt, hauling dirt, and fishing. He was forced to listen to the denunciations of his father, but Tao never wavered in his affection and loyalty toward Zhijun.

Tao was seven when his family resettled in Shanghai and he could finally begin his education. After doing well at the region's best elementary school, he was sent away to one of the nation's most prestigious high schools. After that, he attended Harbin Institue of Technology, where he became a specialist in the control of missiles.

In 1988, Zhijun and Lihui sent Tao to Canada, where he studied at the University of Alberta and then at the University of Toronto. Tao planned on becoming a professor in China or Hong Kong, but in 1993 he tried business, founding a joint-venture company called PanAsia Resources, an investment firm that focused on underground mining, mostly diamond mines, throughout northern China. He caught the eye of Robert Friedland, the well-known international tycoon. Friedland hired Tao to manage his portfolio in China. While working for the investment wing of Friedland's Ivanhoe Mines, Tao invested in one new-economy deal that had been brought to him by Bo when his brother was with Robertson Stephens. Ivanhoe invested $1.5 million in SRS.

When the Asia crisis hit in 1997, Ivanhoe pulled out of many of its Asian investments and Tao left to go out on his own. He bought and sold businesses until, the next year, he met with Yan Yixun. Drs. Yan and Jiang were launching a government-backed venture capital fund that would encourage Chinese technology start-ups. Tao, who had experience in investing and "an entrepreneurial passion," as Yan puts it, was asked to lead the company. The first fund, a test round launched in 1999, was given 118 million RMB, about $22 million of the government's money.

When you see them together, you can't help but notice that Bo looks at his older brother with a degree of reverence. In fact, when I remark on the fact that Tao seems to be following in Bo's footsteps, Bo instantly corrects me, saying that Tao has always been ahead of him. "Tao was the great student, always clear and directed," says Bo. "He was the swan and I was the frog." Still, I point out, he has followed his younger brother into venture capital.

"Well," says Bo. "Maybe, but I was the frog for a long time."

On Tao's desk at New Margin is a *bitong*, the pen vessel for the weasel-hair calligraphy brushes he wields when he has time (rarely these days). Besides being a master calligrapher, Tao is a collector of antique porcelain and Chinese scrolls. Behind his desk is a remarkable scroll with a seventeen-hundred-year-old brushed poem by Wang Xizhi, one of China's most respected calligraphers.

At New Margin, Tao has a staff of thirty people, including a half-dozen partners, analysts, and support staff. Over a relatively short period of time, New Margin has invested in a list of strong companies, including an incubator of early stage e-commerce, Web hosting, and Web design companies, a site devoted to city life and entertainment, an "e-publisher" and literature portal, and two "e-consulting" companies.

When they dine together at a Shanghai Japanese restaurant, Tao notices his brother's weariness and commiserates. "Yes, it's a difficult time," he says. "However, the fundamental assumptions about the technology are valid. Our job is to get the companies strong enough to be here when things take off."

New Margin and Chengwei are competitors on some deals and partners on others, and the men discuss some of their jointly backed companies, including Red Flag Linux and InfoSec, the online security company. Tao is also a backer of Gao Limin's StockStar, the country's most popular site among the nation's rapidly increasing number of traders. At 60 million and increasing fast, China has the largest number of individual investors in the stock market anywhere other than the United States. Many of China's "stir fryers," who get into a stock until it heats up and then quickly get out (staying in somewhat longer than U.S.-style day traders), spend their entire working days on StockStar. Tao quickly earned back New Margin's initial investment with a second round of financing led by Goldman Sachs. In this round of $5 million, which represented 10 percent of the company, the valuation shot up to $50 million. Chengwei was in the round at half a million dollars.

At dinner, Tao is complaining about a CEO. "He's waiting by a stump for a hare," says Tao. In an ancient Chinese story, a farmer starved to death watching a stump in the woods while waiting for a hare to dash into the stump and die. "I believe we may need to bring someone in."

Bo distractedly agrees, but he seems uninterested in pursuing the conversation.

Computers begin "thrashing," or slowing down and malfunctioning, when too much information is being simultaneously uploaded from and downloaded to the microprocessor. When I next see Eric in San Francisco in late July, he says that he is thrashing. We're at the Mandarin Oriental Hotel in San Francisco, which he uses as a living room, after his latest trip to Hong Kong. His normally meticulously combed hair is sticking up. He wears a suit, but his necktie is loosened, and there are dark circles below his eyes. It looks more serious than the normal jet lag and lack of sleep.

Eric explains that he and Bo have had a series of meetings about the direction of Chengwei. He admits that their original strategy was somewhat scattershot: small investments in a lot of companies, since the odds were better that some investments would pay off. (It would only take a few big scores to make their $60 million fund profitable.) The strategy would probably have worked a year or two ago, but not now. "No more small investments," he says. "We'll invest more on fewer companies. We will be completely committed to the companies we choose. We won't let a company die."

Bo, when he arrives, looks wasted, too. He's just in from China for a quick few days. He tries a double espresso, but that won't cure it. He is dragging. The pace and pressure are catching up with them both. He and Eric are together to pore over a new deal. They have high hopes for this one, a company called IEI Technology, a systems integrator in Beijing. Bo says that IEI is a perfect example of the type of companies that should thrive in the current economic climate. Whereas AsiaInfo is the nation's largest systems integrator of telecommunications companies, IEI is fighting it out in the old-economy sector. It would be a major commitment for Chengwei. Chengwei may invest $4 million, and Sutter Hill and Mort Meyerson, whom they're meeting later in the day, may put $2 million apiece. For the combined $8 million, the partners would own 40 percent of IEI. If it goes through, this will be their biggest deal yet. "Bo and I are finally getting comfortable in the driver's seat," Eric says. "This is eight percent of the fund, but it's an ideal partnership with the founders. Our capital can help them become one of the powerhouses of China. In addition to the money, they need our help building a management team who can grow the company."

Bo says that besides more cautious investing overall, the changing climate made them reexamine the type of companies they'll back. Originally they planned to invest 40 or more percent of their fund in dot-coms and the remaining 60 percent in infrastructure companies. Dot-coms are now out. Bo says, "People change clothes with the seasons, but they continue to use the same dry cleaner. The local laundry doesn't care whether their customers are wearing a T-shirt or Armani. The only new investments we'll consider now are the laundries on the Internet."

IEI is a laundry. Its founders, brothers Jack and Steven Liang, aren't the typical youthful entrepreneurs of Chengwei's other investments, either. They are older—"responsible," Eric says—formerly of Sun and McDonnell Douglas, respectively. "They know how to run a profitable company."

At night Bo and I head out on the town. Unfortunately, the Porsche is in the shop. (*Again,* but I don't rub it in.) That means that Bo is behind the wheel of the minivan. (That I do rub in.) It's extremely satisfying to witness. If only his friends in Zhongguancun could see him now. He drives the car in such a way as to tell everyone on the road that this is his *wife's* car. Driving back from our dinner on Polk Street, Bo zooms up Russian Hill on the steepest part of Filbert Street. At the crest, the van shoots into the air and flies for a good ten feet.

Bo is spending less time in the Bay Area; there's just too much going on in China. Indeed, he heads back in a week.

In Shanghai, Bo and Edward meets for a quick dinner, when he tells Edward about a cocktail party planned for tomorrow. He's bringing together most of Chengwei's entrepreneurs and asks, "If you were talking to them, Edward, what advice would you offer about operating in this business environment?"

Edward answers with a word: "Budget."

I arrive in town the next evening and Bo picks me up at my hotel. Jeffery Jiang is in the car, too. Bo was inspired by stories about Richard Branson and says that he wants to be as good of a manager as the Virgin boss, encouraging the loyalty of his employees by treating them with respect and providing them with opportunities. Jeffrey is the first beneficiary. Chengwei plans to send him to Stanford for his M.B.A. A job is promised for when he completes the program. The opportunity is extraordinary, Jeffrey says. "My family are peasants and workers, so this life is unexpected." Another beneficiary of Chengwei's attempt to provide a "new model" for Chinese employers is Ma Ying, Bo and Eric's office manager. When Bo learns that she is working nights and weekends on her M.B.A., he and Eric insist on paying the costs and giving her some time off for schoolwork.

Xu Bing has the cruise control set as we glide over the Hangpu River by way of the twisty columned bridge named after Deng Xiaoping. Yo-Yo Ma's cello plays loudly and profoundly. Now Xu pilots the car over the brand-new and uncrowded Shanghai freeway that leads out of town. It's lined with maples the color of red grapes and pink-blossomed oleanders. We drive by numerous billboards for Internet companies, including Sina and Renren, and enormous numbers of modern apartment buildings on what had been, a year ago, farms. Meant to provide housing for Shanghai's growing number of affluent commuters, the apartment and condominium complexes have swimming pools, tennis courts, satellite TV, and gymnasiums.

Bo jumps out of the car in the shadow of the Mori Building tower in Pudong, the location of the cocktail party where many of the

Chengwei entrepreneurs are gathering. Heading to the meeting, Bo reaches Eric on the cell. Back and forth in English and Shanghainese, they plan the tact Bo should take. "We need new business models for anyone underwater," Bo says. "I guess I just need to drive in the point."

The bar looks out over the shipyards, where an enormous cargo ship is almost ready to be launched. The smoggy sky forms a brown and pink backdrop. In front of it, the lights of Shanghai are being switched on. The city seems painted and Emerald City–like.

About three dozen guests, mostly entrepreneurs, some of whom look as if they haven't seen the world outside of their cubicles for months (the classic signs: vampirishly pale, circles under their eyes, blinking), drink Coke or beer. Uncomfortably dressed up for the party, they wear white shirts and wide neckties, or in some cases sport coats over logo T-shirts. They munch sushi and spicy chicken wings.

Bo, holding a Perrier, tests the sound system at the lectern and then addresses the group. His speech is brief and articulate. He speaks with assuredness, putting the stock market in perspective and reminding the group that it is not about the stock market itself, but about the building of businesses and the building of China. "Gravity is once again a force in our lives," he says. "Can you survive in the real world?" Bo refers to last night's dinner with Edward and says that he asked Tian to advise the Chengwei entrepreneurs. "I expected some lofty, inspiring advice about the Chinese revolution," says Bo. "There's no one who is more committed to this movement than Edward. But Edward said, 'Budget.' Budget. It's the only advice he offered. Where is your money going? From where is it coming?"

He says that Eric will be arriving in China tomorrow and will visit each company to continue to go over their budgets. Bo cites a cautionary tale about Netease, the third-largest portal behind Sina and Sohu. It was worth $400 million before it went public, a valuation it was able to use to recruit and expand its technology team, incentivize employees, advertise and market, and acquire strategic partners. Since the IPO, however, the valuation is down to under $100 million. Employees' options are worthless. There is no money to continue to hire good staff, never mind to invest in R&D, design, and other things that would improve the product. There are rumors that the company is on the block or that it could go bankrupt.

Bo continues, "At this tough time, Eric and I want to refocus with you on your business. We're going to work with you in every way that we can." He then tells everyone to look around the room. "Part of being a Chengwei company means that you aren't alone. Brother companies can work together and help one another. In this group, you may see the marketing specialists you're looking for; the systems people you need; the Web design and hosting services. The exchange of expertise is a valuable tool. It saves cash and allows you to beta test your products and services." Before he's finished, Bo successfully builds a sense of camaraderie. The entrepreneurs do feel less isolated. The call to arms—Budget! TRIM BURN RATE!—isn't personal. It's a rallying cry from a respected general. He concludes, "Yes, it's a different scenario than we expected, but it's a healthy one. Good ideas and good companies are the ones that survive. Your business plans are like gravity. They are what keep you firmly on the Earth." He says something in Chinese. Another in his collection of fish metaphors. It means, "It's better than fish coming on the land." That is, companies that thought they could survive anywhere now have to swim in the ocean. The ocean is the real world where the laws of physics apply. Still, it is better than fish coming onto land, which implies a certain death.

Bo is sitting up in the car, breathing easier again. Nothing's changed, but it may be because he realizes something that was obvious at the meeting: He's good at what he does. In addition, surveying the room at the cocktail party, one can't help but be impressed. He's feeling optimistic again. They just might pull things off.

The car speeds away from the party and gets stuck in traffic under the river and then comes up for air on the Bund. Bo wants to call Eric to tell him the details of the event, but it is 3 A.M. in California and he decides to let his partner sleep. The car drops us on the Bund, and Bo takes my arm and moves me up the stairs at a gallop until we are inside the elegant rooftop restaurant M on the Bund. Particularly overdramatic after a bottle of a good South African wine, Bo is waxing his lofty vision again. He's aware of his wavering mood: "I'm either certain we'll revitalize China or we'll fail miserably. There's no middle ground."

In the morning, I fly to Beijing, which seems to be transformed every time I see it. The new airport terminal is open. There are whitewashed erector set beams that curve over the top of us. There is Internet access from a public terminal in the lobby. The airport is efficient and futuristic.

I hop in a car that scoops up Edward in front of his office and takes us down the center of Beijing on crowded Changan Jie (Long Peace Boulevard). Edward points out the windows. "We laid down fiber there." "We laid down fiber under the subway there." "That building—we're wiring it from the ground up." "See the railroad line? Our wires go under there."

Continuing down the main Beijing thoroughfare, Edward points out the Ministry of Information Industries—"my boss"—and the latest I. M. Pei structure, a bank that looks as if it could fly. Then he indicates an enormous building with a stark mirrored exterior that goes on for several blocks. It's the Oriental Plaza, the model for the world's office buildings in the future, he says. At 8.5 million square feet, it is the largest office complex in Asia. Edward says that he's wiring the entire building with high-bandwidth fiber and connecting everything, from the control center in the basement to the "smart desks," prewired and connected to the Internet as well as the printer, a scanner, and a video-conferencing system in each office. "There is one hundred megabytes per desktop."

The consumer electronics giant Sony has leased a large portion of the lobby of the Oriental for a showroom of its products of the future. The Web-enabled digital products are all based on the premise of a high-bandwidth Web connection, all combining traditional consumer electronics products with digitized movies, music, and images. CNC is working with Sony to provide the bandwidth. "It's a wonderland, a playground for the future," Edward says.

Edward and I drive through evening traffic to meet Bo, who took an afternoon flight from Shanghai. We stop near Bejing's Hou Hai, or Back Lake. Down a *hutong*, we bend to pass underneath a stone arch-way and walk along a path through a bamboo grove. The restaurant's round doorway is lacquered red. Bo is waiting up a flight of stairs in a small private room when we arrive. Somehow fittingly, he's reading Kierkegaard's *The Concept of Dread*. Edward teases him about his choice of reading material. "It should be Lu Xun," says Edward.

The reason is that this Pinganli District restaurant, Kong Yiji, is named after a short story of the same name written by Lu Xun, China's most renowned twentieth-century writer. Is it a coincidence that Lu Xun, the inventor of the modern Chinese short story, was also a translator into Chinese of many of the novels of Jules Verne, author of the book that taught Edward to dream?

Elaine Wu, the shareholder relations liaison at CNC, selected the restaurant. Rarely in China are things chosen randomly. The setting is poignant, a poetic message to Bo and Edward and their foreign guest. In his twenties in the first decade of the 1900s, Lu Xun rejected the backward, superstitious thinking of the Chinese. He believed that only scientific thought would improve life for the Chinese people. He rejected this, too, however, and went so far as to change careers from medicine to writing because he decided that the Chinese were ailing spiritually and intellectually more than physically. He "decided to devote himself to the creation of a literature that would minister to the ailing Chinese psyche," according to the introduction of a collection of Lu Xun's work, *Diary of a Madman and Other Stories,* translated by William A. Lyell. Though he never gave up on that mission, he did change his mind about the problem with China. Writes Lyell, "Instead of blaming China's backwardness on the 'inherited superstitions' of the common people, he now appeared to turn against members of his own class, asserting that the accusation of superstition was only a convenient slogan of hypocritical gentry scholars who hoped to absolve themselves of responsibility for the current national crisis by laying the blame for China's unfortunate plight at the doorstep of the common people." Lu Xun "called upon his fellow countrymen to become 'warriors of the spirit,' writers who would give voice to the sufferings of China's silent masses and articulate their hopes and fears, writers who would, at the same time, exhort the whole Chinese people to reform their society and stir them to resist oppression."

Kong Yiji, whom the restaurant is named for, was an aged Chinese scholar whose erudite learning had grown increasingly irrelevant since China began to modernize. A scholar who spoke with pretension— "When he talked, Kong Yiji always larded whatever he had to say with lo, forsooth, verily, nay and came out with a whole string of such phrases, things that you could half make out, and half couldn't," wrote Lu Xun. Kong Yiji was an anachronism in the modernization of China, a relic: a scholar rooted in Confucianism (and not a very good one).

Denying his irrelevance, Kong Yiji spent his time drinking yellow wine and eating fennel-flavored beans at a Lu Town wine shop, spouting arcane ideas. He was depressed and lethargic, withering away at a time that progress, not Confucius, was the heart and soul of China.

"Elaine is sending a message to the three of us," says Edward.

"In China, we can't just eat," Bo sighs with a laugh.

In the story, Kong Yiji begins stealing in order to maintain his meager lifestyle. At the bar, meanwhile, he is the object of derision, taunted by the other patrons because of his pretentious speech. Then Kong Yiji stops visiting the tavern. Finally the patrons who taunted him learn that he was caught stealing. The punishment: His legs were broken off. Some time later, Kong Yiji does return to the tavern, hobbling on his hands. "[Kong Yiji] kept talking—more or less to himself—but every last bit of what he said was of the lo-forsooth-verily-nay variety that nobody could understand," writes Lu Xun. "At that point everyone roared with laughter, and the space within the shop and the space surrounding the shop swelled with joy."

"Indeed, how are we going to be sure that we bring the entire country along?" asks Edward. "We have to remember history and tradition and remember the people who are not part of the current revolution. We have to know that we are the people who are leaving Kong Yiji behind while at the same time we shall become Kong Yiji, no longer relevant."

This feast includes a salad with cilantro, sliced tofu skins with daikon in vinegar, sweet duck, Chinese squash with bamboo, greens, goose wings, pigs' feet, and a vegetable stew called "like a picture of Spring." There are clams in peppery sauce, slippery white buns filled with sweet bean paste, green beans with flowers, and small black crabs.

After dinner, we walk through the restaurant's garden along the lake. It's after ten, but the summer sky has remnants of an orange sunset that reflects on the black surface of the still lake. When an angler, who wears rolled-up pants and a rag on his head, casts his line, the baited hook breaks the quiet surface of the water.

CHAPTER 16

THE CARP HAS LEAPT THROUGH THE DRAGON'S GATE

In early August, Bo calls my Beijing hotel room in the morning sounding like a man who hasn't slept in a fortnight. When I ask about it, he admits that it was an all-nighter *again,* this time with a bunch of bankers who are in town for a conference. We taxi to CNC, where preparations are under way for the company's first birthday party. Edward, whose suit jacket is hanging on the back of his chair and whose shirt sleeves are tightly rolled up above his elbows, is busy overseeing a meeting. It's about the IP 800 project—the one that allows companies to immediately call back people who visit their website and indicate that they are interested in (or who have questions about) their products or services. Next I sit in when he meets with a representative of the Chinese Construction Bank, CNC's partner in an IP calling card that offers personalized calling plans tailor-made for individuals who frequently call specific phone numbers, time zones, regions in China, or foreign countries. After that, he meets the director of a CNC datacenter with whom he is discussing the Web-hosting operation. Edward explains to me, "It means that an entrepreneur has a big idea in the morning and we can have it up and running on our datacenter that night. It's like oxygen so the Chinese businesses can grow."

Overstuffing the central CNC office, spilling out into doorways and hallways, there's now an all-hands meeting of employees. Though the board originally agreed that there could be an employee stock option plan, it has been complicated to implement in a government-backed venture. Indeed, there have been attempts by some shareholders to rescind the promise. One reason was the fear that the move could be precedent setting. Besides the concerns about the costs, it represented another stage in the transfer of power and assets from the government to the nation's citizens. (Ironically, ownership of companies by their workers is a Communist principle that was one of the first abandoned ideals.) In spite of the opposition, however, Jiang Mianheng and Yan Yixun helped push it through almost a year later than promised. (And it indeed turns out to be precedent setting. After the announcement of the CNC deal, Legend Computer becomes the second government-owned company to offer a stock option plan.)

After throwing on his jacket, Edward, looking dignified in his dark suit and speaking fluidly and powerfully, addresses the staff, explaining the historic accomplishment. Chinese workers for the first time will own shares of their company! "It's your company," he says. "In my mind it always has been your company, but now you are legal owners. Make it something you're proud of."

As he hands the meeting over to an HR manager, an assistant tells Edward that the heads of Sina.com have arrived for the signing ceremony. Indeed, Wang Zhidong is waiting in a rarely used office for visiting board members with a beautiful Ming desk and *kang* and framed armchairs. Zhidong is accompanied by Wang Yan and other representatives of Sina. Zhidong and Edward exchange a long and hearty double handshake.

This ceremony is the result of a dinner that was instigated by Bo in July. Edward and Zhidong knew one another, but Bo encouraged the meeting in the hopes of the formation of a strong alliance between CNC and Sina. It sounds as if it worked.

Edward introduces some staff members to Zhidong and the other Sina representatives and proposes a toast with a tall cup of jasmine tea. He says that the companies will have a long and prosperous relationship. Working together, he says, "we will be creative and cooperative and accomplish truly great things for our company and for China."

During his introductory PowerPoint presentation, Chief Operating Officer Chareleson Zheng, whose eyes are wide and eager, gives an enthusiastic history of CNC and describes its network, "the most advanced fiber network in the world." With satisfaction, Zheng notes that China Telecom tried to develop a similar system but failed. "China Telecom was the sole carrier of information in China so there were no innovations," he says. "CNC is bringing innovation to the telecom world. All sorts of ISPs, private networks for companies, city infoports—everything will be changing. There are an enormous array of business offerings in the works from an ATM network to high-end virtual private network (VPN) tunnels for businesses to hook onto the net. Any company that wants to do something creative with voice, data, and video can do so now because of CNC's bandwidth." It's a presentation geared for youthful revolutionaries who are committed to rocking the status quo. Particularly when he says that his company's success will be measured by how much it weakens China Telecom, the worst example of China's state-owned enterprises. Edward can't resist adding, "The summer holiday is coming and China Telecom heads are taking a two-week vacation with their families." Slowly surveying the room for maximum effect, he says, "Meanwhile, we will dig up another three hundred miles, connecting another city to our network."

Sina's Wang Yan, with a raggy new haircut and a typically bright necktie (lime green), describes ways that Sina can utilize the CNC broadband system. There are plans for an array of new services for its users, including PC-to-phone service and video feeds of live events as well as basic broadband ISP and VPN services. The initial partnership includes a deal that has CNC hosting Sina's coverage of the Summer Olympic Games to be held in Sydney in September. In addition, Sina.com will become the major online distributor of CNC's IP phone card and telephony offerings. He indicates that managers of both companies are exploring new ventures, including Sina Net chats using voice instead of typing.

Zhidong, who is wearing his usual collared work shirt with the Sina logo, reports in on the latest at Sina, and ends the ceremony with a pitch for "a long and nurturing" partnership "based on our mutual respect for one another." Edward adds, "As we go forward, we shall both find more ways of cooperation. Think of CNC first. We will think of Sina first. Together, we can continue to build the world's most vital Internet on the planet." The two Chinese IT stars again shake hands.

On his way out, I congratulate Zhidong about his and Liu Bing's twins, which are due early next year. A baby conceived in the Year of the Dragon and born in the Year of the Dragon is very lucky—a double dragon. Twins make it a double double dragon. His joy is obvious, but he is only half kidding when he groans, *"Two babies!"* and shakes his head.

Zhidong and his crew duck out, and Edward returns to his office. Bo's here, too, accompanied by John Sunn from Oval. They enter Edward's office. John boots up his laptop on a round corner table. Nearby, Bo thanks Edward for agreeing to listen to John's Oval pitch.

Edward sits in front of the laptop. John, who is clearly nervous, indicates the first slide of the PowerPoint presentation on the screen. Bo leans back on a low bookcase along the far wall and Jeffrey Jiang, sitting at the table, takes notes.

"The Oval E Manager would help CNC with our suite of Web-based products that enhance and extend the traditional operation of a company," John begins. Charts on the computer screen indicate how E Manager could save CNC time and costs because of its seamless management of sales, order management, procurement, and logistics. "Our system reduces costs by automating the order procurement system and integrating the system with manufacturers, distributors, and logistics. We help manage inventory and reduce product returns," he says.

Edward quietly interrupts. "How are you different from the solutions offered by Oracle or BroadVision?"

John explains that Oval's system, simpler and cheaper, is Java based and runs not on a PC but on the Web. He repeats, "It's cheaper."

Edward says, "At a company like CNC, so much is at stake that cost isn't necessarily the most important factor. Function, reliability, and scalability are far more important. The proven brands have track records. I don't see what you offer that's different for a company the size of CNC."

John becomes flustered. He fidgets in his chair. "This shows why we are better," he says, indicating a new slide. It repeats what John just said. One bullet point promises a "more efficient operation." It and the other points are so general that John is quickly losing Edward. John tries to think of something new to say, but Edward's patience is waning. When I look at Bo, watching from the corner of the room, I'm surprised to see that he seems amused—even when Edward abruptly stands and says, "Thank you for coming. I have an appointment."

Edward's eyes meet Bo's. Nothing is said.

Edward leaves the befuddled man behind. Bo still doesn't say anything as John packs up his computer and briefcase. John finally looks toward Bo and says, "It was a disaster. And to *Edward Tian!*"

The humbled CEO is ready to leave when Edward peeks back into his office. Before dashing off again, he says, "Next time I see you, have answers to my questions. I know that you will be better prepared. Whether you are presenting to me or anyone else, you must anticipate, *'What is in the customer's mind?'* Before you arrive, you must contemplate, *'What is he thinking?'* Companies like CNC have a huge internal tech team that can distinguish what is different about your software. For us, sales and marketing is a bottleneck, not software. So your pitch must be refined for us. I'm sure you'll do better next time."

Edward is gone, and Bo still hasn't said a word. Nonetheless, as he leaves the room and skips down the hallway, there is a sort of spring in his step; he seems strangely calm. He whispers to me, "John is in awe of Edward. Now he's crushed. He'll never again be unprepared."

A taxi drops Bo and me off at the China Club, a restaurant in the palace of the forty-seventh son of the Emperor Konshin, built four hundred years ago. It's a renovated "square house," or *si heyuan*, a courtyard palace in the traditional Chinese design with *liuli* tile with interlacing lotuses and curling waves lining the sharply peaked roof. Inside, dark wood is framed by lacquered red trim. There are handsome, wood-framed table lamps with fat oval shades. Circle chairs made of *huanghuali* wood are set in the room's four corners, and the center court has a "well to heaven," or skylight garden. Along one side is the Long March Bar, resplendent with lacquered mahogany.

We are meeting Bo's father, Feng Zhijun, who waits for us at a round table in a serene dining room near a wall scroll with a scene of a rushing river and a bamboo grove. Zhijun, with a thick shock of graying hair swept back, wears black: a black shirt with a traditional indigo tunic and a Western suit jacket and pants. In his early sixties, he is powerfully built with a wide carriage, bent slightly over. He wears thick, oversized tortoiseshell glasses.

Zhijun presents me with a copy of his new book, a treatise about developing the western territories of China now that the coastal cities in the east are beginning to thrive. "We're planning it meticulously," Zhijun says. "We have a good start in the increasingly prosperous cities, but most Chinese are still poor and without opportunities. That will change province by province, city by city, village by village." Zhijun speaks as if there is a much larger audience—as if we are in the hall at the People's Congress. "There's already evidence if you look at the changes of spending patterns in China. They indicate growing affluence. People who could afford only a bicycle are now buying a TV or washing machine. Within a generation, some will be able to buy a car and maybe their own housing. Why is the accumulation of material possessions important relative to social change? Their growing wealth is the point. Eventually, a middle class emerges. The emergence of a middle class in China will stabilize the entire society. It's a cycle that continues to strengthen the country. There will be more of a market, and that will encourage more innovation and competition as Chinese buys—by 2003—sixty million cell phones, one hundred million TVs, and fifty million new computers each year. The capital from supplying that market will increasingly go to R&D and new developments. Successful companies will spin off new entrepreneurial ventures. The pervading technology not only raises the standard of living, but contains inherent changes to culture and society."

They sure don't talk about the weather in Bo's family. Bo rubs Zhijun's back throughout much of the dinner. He touches his arms and face. When Zhijun speaks, I learn where Bo gets his passion for language and metaphor. Indeed, whether discussing literature, history, art, or green tea, Zhijun is articulate and opinionated. Like Bo, Zhijun recites poetry and seems always ready with an appropriate quotation, drawing from Mao Zedong, Bill Clinton, Confucius, or the *New York Times*.

After dinner, we continue talking in the breezy China Club court-yard, where Zhijun tells a story. "Emperor Qianlong in 1793 received a letter from King George III. In it, the British monarch asked for an exchange of trade and culture and suggested that delegations from the two countries could visit the other and learn."

As Zhijun talks, Bo takes a last drag of his cigarette and then drops it on the brick courtyard and crushes it with his heel. A sallow moon is sinking behind the moss-covered wall. "Qianlong wrote back and said that the English could send people to China to learn from us, but as for the other way around: 'No thanks.' What was there to learn?" Zhijun smiles. "Things have changed. Now China must learn from the West—good business management and the ability to think freely in order to make the best decisions." He winks. "Of course, we can also learn from your mistakes."

When I ask Zhijun his position on China's entry into the World Trade Organization, he says that he supports it wholeheartedly for social as well as economic arguments. I ask what he thinks of those in America who would deprive China of admittance into the WTO and who each year attack China in the most-favored-nation debate. I explain, "China seems unwilling to discuss some important issues. Human rights. The prison labor. Child labor. Torture in Chinese prisons. Religious repression. Freedom of speech. Americans care deeply about these issues. Isn't it reasonable for the West to insist on conditions for free trade?"

Zhijun says, "I agree with President Clinton. Clinton said, 'We believe that trade is the best way to integrate China further into the family of nations and to secure our interest and ideals.' It is an accurate viewpoint. Nothing has opened China more than the open market. Technology and the inherent economic development from it are the next step. Openness and exchange of ideas come with it. An isolated China has no motivation to change—in the ways that we decide we want to change. It's an important point. Americans' motivation may be laudable, but their tactics sometimes leave something to be desired. No one likes to be told what to do. Instead, let us evolve. Since the Cultural Revolution there have been attempts to change China that failed. However, Deng Xiaoping opened up the door to a force that is chang-ing China at a natural pace. To do anything other than work for change in this way is irresponsible, letting down the people who will benefit from an open, successful China. To shut the door on economic change is to turn your back on one billion Chinese." By the time our evening together is over, I have come to understand that Bo's decision to return to China out of a sense of duty and patriotism was probably bred long before the Tiananmen Square massacre.

Bo and Zhijun hold hands as they walk out of the restaurant. After tucking Zhijun into a cab, Bo says, "The older I get, the more I realize that he's in me and I'm in him. A son cannot help but carry his father everywhere he goes." Bo talks for the first time about how demoralized Zhijun was after Tiananmen Square. "My dad goes on," Bo says, "but he never recovered from that time. He contributes what he can, but he would have been a great progressive leader if he had not been marginalized."

Bo once again is exhausted, and it seems as if he'll finally turn in early and call it a night. There's another meeting with yet another Chengwei company scheduled for 10 P.M., but I suggest that he postpone it until tomorrow. However, I watch how Bo deals with the intolerable physical limitations when they arise. He shakes off his exhaustion. Literally. Physically. He shivers, stretches, and stuffs away the tiredness. He stands up straight, smacks his hands together, and lets out a whoop. "Let's go get a drink."

A rattling Beijing cab takes us to the Courtyard, a restaurant and bar off a cobblestone street lit by a weak yellow streetlight alongside the Forbidden City. The large courtyard building, like the one in *Raise the Red Lantern*, must have been the residence of a rich and noble family. We skip through the crowded downstairs, with tables set with white linen and fishbowl wine goblets, and climb up a set of stairs to the cigar lounge in the attic. We sit on plush leather couches behind a rickety Tibetan table. A line of Laughing Buddhas size us up from a high shelf. Through small windows, we can see the moat filled with murky water that laps the Forbidden City's stone wall. Beyond is the upturned golden roof of the Hall of Supreme Harmony. The sky is extraordinary clear and pale blue. It makes way for another spectacular sunset, with flame-colored bolts streaking the sky, a stunning backdrop to the gray bricks and faded red of the emperor's palace walls. A man below us works on bicycles at a bicycle-mending stand. A few children run along the riverbank.

Any sign of Bo's former weariness is gone. He's animated, pacing with an excitement that's reminiscent of the children outside. "Look at that! How beautiful. I wish Tiger could see this. I can't wait until he is old enough to understand this history, this time in history, this life we are living."

After a cup of wine, we leave the old place under a milky night sky. Bo signals a taxi and we're off again. He drops me off at my hotel and goes to his meeting.

Six hours later, in the early morning, we eat a crunchy *youtiao* on the run. The car speeds across Beijing to Legend Computer's headquarters, where Bo is expected by Liu Chuanzhi, the CEO and chairman of what has become the number one computer company in Asia. The Legend story is a great inspiration to other Chinese technology companies. No one inside or outside of China, including Liu, thought that a domestic company could compete with global brands such as IBM and Compaq. However, Liu Chuanzhi has surpassed them all.

Legend Computer was an unlikely offshoot of the Chinese Academy of Sciences in 1984. The academy had been devoted to researching the most advanced technology in the world, but there was no consideration of commercialization. In fact, commercial research, viewed as antithetical to the mission of the institute, was looked at with some disdain. "At the academy and throughout China, the only scientists who would go into commercial ventures were inferior," Liu says.

That view began to change with Deng's reforms. Liu Chuanzhi was charged with creating a for-profit spin-off of the CAS. It was such a daunting assignment that Liu became depressed, certain that he would be unable to transfer his knowledge to the commercial realm. It caused him to question "the meaning of my existence as a human being," he says. He felt that a commercial venture was counter to his training and beliefs; if this was the direction his government required, "life was false!"

Without a choice, however, he tried. With only 200,000 RMB (about $25,000) from the chief of the Institute of Computer Technology, a division of CAS, he set out to create a personal computer company that could compete with the foreign brands that had begun to be imported into China.

At the Legend office, we sit at a conference table. Liu, who is thin and has an angular face, sits near his assistant, who wears pink and has a Barbie watch. After small talk, Liu tells me that when he founded Legend with eleven computer scientists and a meager budget he expected to fail. He says, "Perhaps I was meant to fail so the academy could prove that we should stick to research."

Since the budget couldn't sustain the company, Liu's first task was to figure out a source of income, so the new company began distributing foreign computers by AST and IBM. Meanwhile Liu studied the management of IT companies in Japan and the United States. Managing Legend, he tried to apply what he viewed as the best of both systems— the family concept in Japanese companies and the American system that encouraged individual thinking.

As Legend became profitable selling foreign-made PCs, his engineers designed their own computers. Rather than modifying Western-made systems for Chinese users, Legend created products designed specifically for the local market, including Chinese-language programs. In addition, it successfully built high-powered but low-priced Intel processor clones.

Legend's cheap but well-designed and reliable PCs soon became the company's core business. By mid-2000, Legend had a growing 30 percent of the market share of all PC sales in China. In the process of creating a thriving company, Legend taught the government a lesson. "In the past we only knew how to convert money into technology," Liu says. "At Legend, we learned that we could convert technology into money." Indeed, the success at Legend led to other bold ventures on the part of the government. The same people responsible for Legend eventually were ardent supporters of Edward Tian's CNC.

In the conference room, Liu shows off the latest Legend product, an Internet PC. A single button connects the computer to the Internet and loads a browser. The cost: 6,000 RMB, or about $800, which includes a year of unlimited Internet service. There is a back order of eighty thousand units a month. The computer has a second hidden business model. It connects new users directly to Legend's portal, fm365.com, which is a popular site in China.

When Bo asks about Legend's Internet service provider, Liu says the expected: It's China Telecom. Bo makes a pitch for CNC. "If Legend is the first PC company to offer broadband service, it will be a big selling point," he says. "Edward Tian's network is ready and already has extraordinary offerings, including IP telephony. A world of multimedia offerings are coming." Liu seems very interested and asks Bo to set up a meeting with Edward.

We leave Legend for CNC to tell Edward. When we arrive and peek into his office, Edward is on the phone. He waves us in. The side of the conversation we hear is tense. Uncharacteristically, Edward seems ready to blow up. When he hangs up, he slams the phone and shakes his head. Without telling us who he's talking to, he begins a tirade of complaints about the time it takes to manage his board. "It's the most tedious part of the job," he says. "I never wanted to be a politician. I must change my face for these people, and I don't like what that feels like. Still they question my loyalty! I try to understand their agenda. I try to make them feel as if they are part of what we are doing, but sometimes . . ." He trails off.

"At AsiaInfo, my three priorities were corporate strategy and execution, dealing with customers, and managing our corporate culture. At CNC, managing the shareholders is first. Next is strategy and execution and the company's culture. It's not the best system by any means." He says that he does his best to follow Feng Zhijun's advice. He relates that Bo's father once counseled him to proceed as if he is swimming underwater—"quiet, humble, keeping below the surface."

Bo says that he doesn't understand why the directors do anything but celebrate Edward's accomplishments. "In less than a year, CNC has increased its valuation a hundred times," he says. "The initial investments of five million dollars by each shareholder company is worth nearly five hundred million dollars. The ministries and government agencies are losing money. With this investment, they have made more than they have made in thirty years."

The tough part, Edward says, is that the people he deals with "grew up in central planning mentality." He resolves himself to be honest with them and respectful, though he is incapable of being obsequious. "It's all I can do." The problem, he continues, is that "we're dealing in compressed time. If you can't handle the compression, you explode. It makes them very nervous. We're preparing for the WTO, preparing for competition, preparing for the new generation of China as quickly as is humanly possible. Is it fast enough? We don't know. Resistance is destructive, which is why I worry about the slow education process."

When Bo tells Edward about Liu Chuanzhi's interest in CNC, his mood changes. "I would be very happy to work with him," says Edward. "He is a great man." He thanks Bo for initiating the conversation and promises to follow up. We say good-bye to Edward after planning a meeting in San Francisco. Bo and I taxi to Beijing Airport, where we are faced with a two-hour delay before our flight to Shanghai. In the lounge, Bo finally sleeps.

When we hit central Shanghai, Bo leaves me on the Bund and heads out into the night with Baojun and another friend for a mahjongg game. It goes all night, so in the morning, when we meet again, Bo is bleary eyed. There's not even time for coffee, though. He's in a rush because of an important meeting across town. Bo calls Xu, the driver, for the fourth time. Xu is apparently racing as quickly as possible in horrendous traffic. We finally see him coming—careening toward us, full steam, the horn blasting, his head poking out of the window so that he can be heard shouting at pedestrians to move out of the way. The A6 screeches to a stop in front of us and we pile in. All the rushing and we're stuck in traffic again, this time under the tunnel that leads to Pudong. (We could breeze through on Shanghai's new subway that connects both sides of the Huangpu, but Bo has never considered taking it.)

Bo has a crazy schedule of follow-up meetings with entrepreneurs set for every couple hours throughout the day. The first is at the Jinmao, which pierces the sky like a scalpel. It's impressive every time I arrive here. We blast up the elevator to the Hyatt lobby on the fifty-third floor of the eighty-eight-story building and run to the restaurant, where an entrepreneur is waiting. Next, Bo dashes off to meet with the founder of a company called iTom, a start-up that is making good money designing dynamic websites for an impressive list of clients, including Sony and Nippon Paint. (For the paint corporation, iTom has created an interactive site at which customers can try out combinations of paint colors on virtual replicas of their homes.) Chengwei has invested $1 million in the company. Next Bo races to Oval Technologies, where John Sunn is busy in front of his computer.

John wants to talk to Bo about the problem he is having hiring talented people, particularly managers and engineers. "Since you stole my CEO, you owe me your help on this," John says, referring to Chengwei's hiring of Yangdong Shao. The problem in China, far more than in the United States, is that managers are at a premium. It goes back to the lack of management training. Indeed, the best-trained managers still come from the United States. To hire students fresh from Harvard, Stanford, or other M.B.A. programs, John and other CEOs have to compete with established American companies—everything from investment banks to international corporations. It was easier to land them during the dot-com boom, since start-ups were making millionaires overnight. Now start-ups are viewed with caution, and the risk is too high for many young business school graduates.

There's a similar problem hiring engineers. He says that U.S.-trained engineers are at a premium and have unrealistic expectations of the market. Chinese-trained engineers, on the other hand, have what John describes as "an inflated sense of their value to companies." He explains, "They come in and don't know how little they know. The best graduates from Chinese universities come in boasting that they have coded five thousand lines of Java in a single year. At Stanford we had to do twelve thousand a month. The gap is startling. Chinese engineers are unprepared for what is expected of them."

In spite of their lack of training, young Chinese programmers are being lured by companies that can offer relatively high salaries and a range of benefits. "They haven't gotten the fact that the free lunch period is over," says John. "One typical interviewee said that she wanted a hundred thousand RMB a month and a housing allowance," he says. Meanwhile, the Oval founders still aren't taking salaries.

The problem in part is a reflection of John's wise decision to avoid trying to replicate a highfalutin U.S. start-up. Instead, the company I see reflects the reality of Oval's position as an early stage start-up: committed but hungry. John weeds out many potential recruits when he tells them, "Come aboard if you're here for the long haul because you want to create a great company. Pitch a tent, don't ask about pay. Let's get to work." It means that the ones who sign on are intensely committed.

John isn't complaining, but he reflects that this down-and-dirty operation is a far cry from his last job with Merrill Lynch. He says that he even has to buy and then hide soap and paper towels for the shared rest room in the loftlike building. Otherwise, someone always steals it. "In a way it sucks," he says, "but it makes you tougher."

Bo promises to help do some recruiting and says, "Things will change once the market downturn sinks in. People will be looking for solid companies and good managers."

"I'm keeping an eye for the chance to pick up gold from the garbage," John continues. That is, he's looking out for qualified people who are leaving companies that implode with the Internet balloon. "There are a lot of good people whose stock options are underwater," he says. "We're here, waiting for anyone who is excellent."

Coming from Manhattan, John was unprepared for another problem that plagues the entrepreneurs in China. The nation's archaic, convoluted, and inconsistent business systems with volumes of red tape can be overwhelming. "Regulations are crazy," John says. He wanted to buy a router for $3,000, but he had to get approval in order to withdraw the money from a bank. John says, "China has a long way to go before it's as easy to do business here as in the U.S., and at the same time we have to compete with the U.S." There are other examples. In some cases companies cannot directly hire employees, for instance. They must work for a government intermediary. However, the WTO is motivating the nation to rework the inefficiencies as quickly as possible.

Bo promises to help John establish a relationship with a banker who can make it easier for Oval to have access to its own cash. In addition, he has a list of some potential Oval customers and promises to set up meetings. As we leave, Bo says, "Let's talk at least weekly. We'll help in whatever ways we can. For now, the big challenge with every start-up in China is keeping your ball on the fairway. As long as you don't go out of bounds, you'll be a player. You're doing exactly what you need to do."

Outside, en route to the next session, Bo says, "Eric and I have to remember to give time to Oval and the other strong companies. The pull is always to try to save companies in trouble, but VCs must work hard to make their good companies even stronger. It's counterintuitive, since every day at a troubled company is a crisis and you feel like you have to do something, quick."

The day progresses and melts into evening. More meetings. "Where are your customers?" he asks founders. "Who have you visited this week? Who are you visiting next week?" It's work he never imagined himself doing. Meanwhile, Bo and Eric are applying some of the same thinking to their own start-up—to Chengwei. "We're telling our entrepreneurs to get some traction before they move on. We need to, too. The only way for us to get traction is to help our companies get traction." Hence: "We will do whatever we have to to get these guys profitable."

CHAPTER 17

SLEEPLESS IN SHANGHAI

The first Chengwei conference is held at the Grand Hyatt on the fifty-third floor of the Jinmao Tower in September 2000. I'm on the same transpacific flight as Len Baker, the keynote speaker, plus Heidi, Tiger, and Lihui. The trip seems quicker than normal because of the distraction of some playtime with Tiger, a veteran traveler at two. He's bouncy and excited when we land, anticipating the sight of his father. When we see Bo after clearing customs, Tiger squeals and Bo grabs him and throws him into the air. Len and I are placed in the care of three young Chengwei assistants while Bo, with Tiger wrapped tightly around him, leaves with Heidi and Lihui.

In the morning Bo shows up in a new and handsome suit with navy pinstripes, looking excited and confident. The many conference goers hover around tables set with urns of coffee and platters of pastries. There's a mix of executives from most of the dozen Chengwei investments and leaders of other Chinese Internet companies, some of them being considered by Chengwei and others hopeful of being considered. In addition, there are analysts, representatives of the Chengwei investors (including the Yale Endowment and Power Corporation of Canada), the press, and assorted other colleagues.

Baker, perpetually boyish in a blue blazer, starts things out with a motivational and inspiring speech. Next comes Edward Tian, who is a celebrity in this crowd. Bo and Eric asked Edward to drive home a point, which he does. He counsels companies to streamline their operations, make and stick to conservative budgets, cut their burn rate, and get paying customers. However, he can't help launching into his usual impassioned conclusion. "History rarely gives an opportunity to change the world," he says. "It doesn't come along even once in a generation. Now we have the chance to make a difference in China after more than a century of stagnation. Yes, it requires hard work. Yes, the obstacles are huge. Yes, we don't know the outcome. But a revolution requires nothing less and make no mistake about it: *We are making a revolution.*" The applause is thunderous.

The entrepreneurs gather for lunch, but Edward is whisked away by Eric, who has a group of Chengwei staffers waiting with a presentation about the newly organized Chengwei Foundation, a group he and Bo started that is devoted to one of Edward's dreams: wiring China's schools. Edward offers his full support, even indicating that he will assign a full-time person at CNC to work with the project. However, he wants to make certain that the foundation has a careful, thought-out plan. "In the U.S., Microsoft and others have found that computers in schools are not enough," he says. "Teachers have to understand how to use them to help their students. Software must be developed to address the specific problems facing educators."

Edward has a plane to catch. He's going to Hong Kong, Singapore, and the U.S. West and East Coasts on a road show to raise a significant round of financing for CNC. Bo walks him out, thanking him for fitting the conference into his crazed schedule.

Over the course of the next forty-eight hours, the Chengwei entrepreneurs present their stories. They range from eloquent and impressive (InfoSec, Red Flag) to monotonic and unfocused (T2 and iTom). Afterwards, an analyst whispers, "There are more good ideas and more energy to carry them out in this room than at all of the state-owned companies combined. Anyone who worries about China's future need only check into this group." I notice how far some of the entrepreneurs have come under Bo and Eric's tutelage. Men who a couple months ago seemed unable to articulate their business models with any enthusiasm now seem professional, direct, and impassioned. In fact, most of them have convinced me that they just might pull off their respective businesses.

Throughout the conference, I am aware, and struck, that Len Baker listens intently to every presentation, taking fastidious notes. Another banker of Baker's stature would have flown in, presented his keynote speech, and headed out. It can't just be his investment in Chengwei, which is substantial for Bo and Eric but trivial for a company the size of Sutter Hill. The only explanation is his interest in participating in the revivification of China and, moreover, his devotion to Eric and Bo.

Though Baker is a backer of Palm and carries a Palm III, he writes on a faded, dog-eared notebook with a spiral wire binding that's unraveling. One page has a scribbled chart with three columns. The first column is headed with a plus sign. Next is a question mark. The third column has a negative sign. Though Baker's detailed notes include a dissection of what he hears in the presentations, this page is a synthesis of his impressions of Chengwei's stable. After the final presenter, he shares his capsule reviews with Bo. "Based on what I heard, I would personally invest in five of these companies," he says. "I would consider three others. I would pass on three others. All in all, you should feel proud." Bo fixates on the companies in the negative column. Sensing it, Len explains a relevant part of Sutter Hill's history. "Out of every ten companies in a portfolio, particularly early stage companies, we expect that three will be successful," he says. "We also expect that two or three won't survive and we'll lose our money."

"What about the other four or five?"

"We'll make our money back, but that's it. However, the good returns come from the fact that we may make ten times or more our money in the three successful companies. If you do that, you grind out a thirty percent return, which is a pretty good return for your investors." Bo starts to say something, but Len isn't finished. "However, we don't typically just pay a thirty percent return. Last year Sutter Hill had a seven hundred percent return. How? One of the companies that succeeds might make not ten but twenty-five times our money. One might make one hundred times. You don't count on that happening, but it can and it does. The point is: You don't need to bat a thousand. You don't expect to. So listen to me when I say that you're looking pretty good. Remember: The company that brings back the big returns does far more than make you and your investors some money. If you create one Cisco, one Palm, one EDS, one HP—well, you have a chance to do that. If you do, you will have created a company that will become a bellwether in China or even in the world."

Bo is thoughtful when he leaves Len. It's as if he is trying to decide if he can believe him.

The conference is over and most of the businessmen disperse: to Beijing, Hangzhou, Hong Kong, Taiwan, London via Moscow, and the United States. Len remains in Shanghai before he flies to Beijing for another conference. Bo, with Heidi and Tiger, will meet Eric and Len in Beijing tomorrow, but this evening he takes the Chengwei staff out for a dinner of thanks for their work on the conference. They celebrate at a renowned and pricey restaurant that specializes in shark-fin soup. When he surfed, Bo felt that it was pressing his karma to eat shark-fin soup, but golfing has no such downside. At the feast, Bo toasts each staff member. "This is a significant milestone," he says. "It never would have happened without each of you. We all will be able to tell our children that we helped usher in the information age in China."

I meet Bo at his apartment in the morning. He's on the phone to Sandy Robertson, who is home in California. I can hear the pride in Bo's voice as he describes the conference. When he hangs up, he seems emotional. Sandy must be pleased.

In Bo and Heidi's kitchen, Shen Baojun, dressed in all black silk, plays with Tiger on the newly waxed floor, throwing around a Ping-Pong ball and crashing a windup Mickey Mouse. Tiger is in stitches. "More, Baojun! More, Baojun! More, Baojun!"

The sound of Lihui's vigorous chopping comes from the kitchen, and soon she is pushing plates of steaming food on us: slivers of tofu and cucumber salad; chicken soup, which she ladles into porcelain bowls; long beans with pork; and green onion flatcakes served with tiny ceramic dishes of chili oil. After the feast, Bo and the family, with suitcases and toys, pile into the elevator. Downstairs, Xu holds open the door for them to pack into the Audi, which dashes out of the courtyard into a torrential rainstorm. Later I'm told that the storm is the trailing edge of a monsoon that is hitting Asia tonight.

Edward, on the road show, finds that the still sinking stock market hasn't dampened investors' interest in CNC. It doesn't hurt that Edward is selling an investment, China's new telecom, that is being run by the world's number one entrepreneur, at least according to the September *Red Herring*. In the magazine's annual list of Top Ten Entrepreneurs, "our heroes . . . responsible for almost all innovation in business," Edward is number one in the world.

During his presentations, Edward is asked about the implications of the WTO on CNC. As far as he's concerned, there's only upside. IT players who come into China will require the breadth of CNC's services, including its broadband network, datacenters, and more. What about foreign competitors? "It's unlikely that foreign companies will be able to catch up to CNC's head start in terms of its infrastructure, so potential competitors will more likely become partners," he says. "No one else will get the right of way to dig up the land to lay down cable. We got that free from our shareholders. Who can compete with that?"

The trip is a success. Edward easily raises $300 million from an influential group of international investors that includes Rupert Murdoch, Goldman Sachs, Michael Dell of Dell Computer, and the Kwok brothers of Sun Hung Kai Properties in Hong Kong. The deal values CNC at $2 billion. The *Wall Street Journal* sums, "The Chinese company arranged the new financing even as panicky investors dumped holdings in tech companies around the world." The *Journal* quotes Thomas Ng, head of Venture TDF, a Singapore-based VC fund with ties to the Shanghai municipal government. "[CNC is] taking the approach of a WorldCom Inc. and taking the cream of the telecom business. They may well become one of the largest companies in the world." The *Journal* article also notes that CNC could have "raised multiples of the money they actually took. They know if they wait, they can get a better valuation."

When he's again in China, Edward seems nonplussed when shown a copy of the latest Asia *Wall Street Journal* in which Leslie Chang has written a front-page piece about new government restrictions on foreign investment in telecommunications companies, including Internet companies. Chang's piece is based on drafts of two new sets of laws and regulations. They come, of course, from the ministry of Wu Jichuan.

Chang writes that documents leaked to her reveal "a torrent of licensing obstacles that could complicate existing operations and even deem them illegal in their current form." They represent the latest swing to the right in the form of increased control and more barriers. First, anyone doing business in the sector needs to be approved by the state. Telecom joint ventures have to have their board chairmen and general managers appointed by the Chinese side. Changes in business plans need to be approved by the ministry. In addition, "activities in violation of the regulations include usage of telecom networks to transmit information that endangers national security, disrupts social order or contains pornography." There are new limits on the minimum revenues of companies that invest in Chinese Internet companies—$10 billion—"which would rule out almost every company operating in China's Internet sector."

Edward dismisses the news with a wave of his hand. The regulations won't effect CNC, he says. When I ask Wang Zhidong about them, he shrugs them off, too. It reminds me of a comment I read from a businessman who runs a restaurant in Beijing. "We do what we want and then find a way to make the regulations support it," he said. "If we have to make adjustments, we make adjustments. In China, it is better to ask for forgiveness than permission." Of course, this type of business isn't necessarily conducive to restful sleep, but, as Bo has said, "in a game of cards, you play the house rules."

Meanwhile, there's encouraging news about the WTO. The United States has torn down more barriers to normal trading relations. Contradicting the tightening atmosphere coming from Minister Wu, the steps toward entry into the WTO include the end of twenty-year-old restrictive trade policies that have kept many U.S. companies from entering China. Now, as the *New York Times* reports, "a new wave of international IT companies, including semiconductor manufacturers and software makers, are preparing to participate in what everyone assumes will be the coming high-tech economic boom here."

In September 2001, in Washington, Congress votes by a wide margin to grant China permanent normal trade relations. The implications for U.S. companies? Tariffs on U.S. high-tech imports will be permanently eliminated, and China could quickly become the biggest market for many IT companies who are already selling more of their computers, cellular telephones, routers, fiber, and software to China than anywhere other than the United States. Also, China promises to protect intellectual property and patents, though there's more to negotiate on this enormous problem. Finally, China has agreed to lift quotas on investments in its state-controlled industries and to open its markets to WTO member nations.

Bo and Eric are back at work on their portfolio companies, paying particular attention to the ones that have failed to sign up customers. There's a critical make-'em-or-break-'em tone to their days. Crisis thinking sets the stage for some radical solutions to help the most troubled ventures. For example, SeeWAP is struggling in what Bo and Eric fear is a hopeless arena. Brilliant engineers are working on wireless applications, but who will buy them? There are only a few possible customers, including China Mobile. However, after developing initial products for nine months, the SeeWAP team has learned that China Mobile has started its own subsidiary to do exactly what SeeWAP is doing. China Mobile represents 60 percent of the company's potential market. Boom. It's gone. In a meeting, Eric is tough. "You've got to stop fucking around," he says. "If sixty percent of your market is gone before you've started, it's time to rethink your business."

The SeeWAP founder says, "It's all right. We're already planning for that. We're going to create killer aps for mobile devices. We won't make money for a long time, but it will happen."

"No no no," says Eric. "No more! *There's no time!*"

He looks closely at the man and says, "We don't want to see you until you figure out a new strategy. Take a look at your technical resources. Examine your core competencies. See what market you can apply your core competencies to in a scalable way. *And go get a customer.*" The man disappears. There's no news for a week. For two. For six. When Bo or Eric call, they're told, "We need more time." Then Bo gets a call that the SeeWAP chief is heading over to see him. (Eric is in San Francisco working with IEI.) Bo expects more pleading for time, but the man arrives with a cat-that-swallowed-the-canary grin. Before sitting down opposite Bo, he says, "I've got Hangzhou Telecom."

Bo can't believe he's serious. "Hangzhou Telecom?"

The man nods. He finally explains, "I kept looking for WAP customers. I was thinking, *Shit, I can't get any!* What am I going to do? Then I did what Eric said I should do: analyze our core competencies. I stumbled on another application for our technology. The companies doing broadband like Edward Tian's CNC and China Telecom are bringing their systems to localities, but then they have to rely on the local city or provincial cable or telephone companies for the last-mile connections. There are hundreds of those last-mile operators that have to provide high-level connections, billing, servicing of customers, and more. It's the same stuff we were going to apply to wireless companies. It's a highly fragmented marketplace. The decision-making power is local, so we don't run into the problem in the wireless industry, where there are just a few centralized providers like China Mobile. Each city or province needs to build infrastructure to link the broadband Internet to customers, whether DSL, cable, or direct fiber. How do they do it? How do they do billing? So we figured this out and rewrote the pitch. The first place we tried was Hangzhou, since it is so far along with its broadband system because of its partnership with CNC."

"How big is the contract?" Bo asks.

"One million. With the promise of more," he says. "Oh yeah, and we probably have another contract. Almost signed. The same type of deal. Shanghai Telecom. Also a million." The man says, "We're no longer a WAP company, so we have a new name." It's OneWave.

Bo calls Eric. They are both stunned.

Another problem company is T2. As Eric and Bo see it, part of the dilemma is that the company's founder refuses to try to get customers. He seems incapable of pitching his product. When they meet, Eric's obduracy is met by a blank stare. "You have to get a customer," Eric says. "*You have to.*"

The man says that he'll try.

Eric throws up his hands. "You had better do more than try," he says. "Let's make a list of potential customers."

They do. The man says that he'll pursue them, but Eric leaves without much confidence.

In a week, he and Bo return together. "Have you signed a customer?"

"No."

"Who have you visited?"

The man looks down. "I've been working on the software. I'll show you!"

Eric is nearly apoplectic. "Your software may bring peace to humanity, but no one will ever see it if you have no business! What do you want me to do?"

Before it's over, the man again agrees to try. However, the next week, the same meeting replays. Eric and Bo leave in despair.

Another investment, iTom, the website design company, has run into a different type of wall. Led by a couple of talented engineers with good reputations and contacts, iTom builds complex interactive websites. For a change, they have an impressive list of customers. The problem with iTom is that it has no depth of services. That is, it builds a company's interactive website, sends a bill, and moves on to the next customer. There is nothing more iTom can do for its customers. In fact, when Nippon Paint requested more services, iTom, a six-person operation with neither the R&D resources nor technical know-how, had to turn down the business. That means that the heads of the company have to continually track down new customers in need of interactive websites instead of focusing on their products and services. It's a model that won't scale.

Coincidentally, Bo and Eric's junior partner, Yangdong Shao, meets a man named Darwin Tu, a Stanford Ph.D. and respected senior manager with Digital Impact, a leading e-marketing software company in the United States. He tells Yangdong that he wants to return to China, so Yangdong suggests a meeting with Eric.

Tu and Eric meet for tea at the usual spot, in the lobby of the Mandarin Oriental Hotel in San Francisco. After they talk for an hour or so, Eric's brain begins cooking. Afterward, he huddles with Bo. "This guy knows what he's talking about in terms of e-marketing," he says. "He has managed a pretty big team with enormous success. He wants to come to China. OK: T2 has technological geniuses but no management that can get customers. iTom has customers but can't provide products or services that will keep them."

The two partners agree that they will propose that Tu, T2, and iTom join up. The synergy seems ideal. iTom could continue to do what it has successfully done in the past: sign customers and build them interactive websites. With T2's technology, the new company could provide e-marketing services for the sites, tracking and profiling customers for their clients. That is, they could build websites and provide ongoing marketing services.

When he hears the pitch, Tu is interested, but the founders of iTom and T2 are reluctant to give up their individual ventures. Eric and Bo work on them, hammering on the precariousness of their businesses. "Here's a chance to take two dying companies and create a new entity: one that could become a lion," Bo says.

The founders relent, but negotiations are complex. Both the T2 and iTom founders feel that their company's valuation should be higher and that they should have more equity in the new company.

"No!" Eric tells them. "You obviously don't get it. We're *getting rid of* T2, *getting rid of* iTom. They don't exist anymore. We are giving birth to a new company. This isn't a GM and Hughes Electronics merger. Give me a break!"

He shakes his head when he hears their counterarguments. "Listen," Eric says, "if we do this, everyone gets an equal third. That's it. I'm afraid you have to take it or leave it. We're not negotiating this any more."

They take it. Chengwei merges the companies into a new one called HDT Incorporated and invests an additional half million dollars. Darwin Tu takes charge as CEO. Bo and Eric are relieved, particularly when the first customers sign on, including iTom's original client, Nippon Paint. HDT takes over the website that iTom had created and provides ongoing data mining and analysis. There's a revenue stream!

Bo and Eric remain frustrated with some of their other investments. Bo once said that he and Eric wouldn't let any of their companies die, but two of them—Rebound, the excess inventory site, and Renren, the web community modeled after Geocities—may. The partners have learned their lessons, though. Chengwei was a minority investor in both companies. As Bo and Eric discovered, minority partners have almost no clout. The companies made mistakes along the way that Bo and Eric were powerless to address. As a result, Rebound may be sold at a bargain-basement price, but it could go bankrupt. If Renren can't be sold, it, too, will go bankrupt.

Meanwhile the IEI deal is on hold. The challenge for it now is to grow quickly and stake out a place as a leader in the crowded industry. To do so, it must increase its revenues from $10 to at least $50 million within a year. The founders admit that they don't have the skills that are required for the company to grow much bigger, and they have agreed to bring in a new CEO. After agreeing to hire one, however, the founders have balked at going forward. The result: time is passing by and IEI is in a holding pattern. The problem has escalated into a crisis, with Bo and Eric battling with the IEI chiefs. Both sides are indignant and angry. Chengwei is ready to invest, but the deal is put on hold.

Overall, the Chengwei portfolio is a patchwork, but more companies are showing promise than not. Bo and Eric are increasingly weary without time to recover, dashing from one troubled company to another. Len Baker reminds them, sometimes on a daily basis, that they are doing exactly what's required. He says, "The suspense you feel with Chengwei—wanting to know how the story is going to turn out—wanting to know if they're going to make money or not—wanting to know if they are going to fail or not—where the biggest successes are going to come from—is the essence of the process. You can't rush things. You can only do the work. If you do it right, there's a good chance of success. Greatness is the result, but the process is what gets you there. Sustainable revolutions come day by day."

"Paper cannot wrap up a fire." The fire burns. The Chinese government attempts to contain it.

In November 2000, the United States passes its version of a trade agreement with China that firmly opens the doors to foreign ownership of up to 50 percent of Internet companies and 49 percent of Internet service providers. The headline in the New York Times reads, "China's cyberspace is officially open."

But not quite. Soon after the trade agreement is finalized, Minister Wu once more raises his head, once again about content on the Net. He says that the MII will enforce the nation's ban on foreign news online. Until now, with the haziness of the regulations as justification, Sina has been carrying foreign news fed them from Agence France-Presse, Reuters, and other Western sources. After Wu's announcement, however, all the main portals instantly drop nonfinancial foreign news and kill links to the websites of foreign news organizations. While Sina's U.S., Hong Kong, and Taiwanese sites are unaffected, Sina in China exercises particular caution about stories it perceives to be sensitive, such as when a man blows himself up with a bomb in Tiananmen Square. The website doesn't run the story until ten minutes after Beijing TV does.

That week there is another worrying report that local governments throughout China are establishing Internet surveillance units and tracking down more people who use the Web to express political dissent "or other nonconformist ideas." As proof, another political website is shut down, and the operators of the site are charged with posting "counterrevolutionary content." According to the New York–based Human Rights in China group, police are searching for the website's organizers.

As expected, there are the usual criticisms in the West, where human rights groups rightfully condemn any attack on free speech and human rights violations. However, there's another view in China, where pragmatism generally seems more prudent than antagonism. At Sina, Yan Yanchou explains, "Caution is wise because so much is at stake. We are willing to work with the government to open things up slowly." Bo is sometimes annoyed by what he views as knee-jerk Western attitudes that fail to look at the complexities in China or don't acknowledge how far China has come in a very short amount of time. "If this much is out there now," Bo says, "imagine what will come next? Give China time." Edward says, "It has only been twenty years since China came out of the Dark Ages. It has taken the Western nations hundreds of years to develop their social and governmental systems."

If technology indeed brings with it a brighter future, as Bo, Edward, and many other Internet entrepreneurs believe, there's every reason to be optimistic. Whereas the Chinese government actively seeks to squash almost every force behind change, it is the Internet's biggest booster. Yes, Minister Wu announces new regulations and restrictions, but most are never enforced, while others are unenforceable. There are tragic arrests, but the government is committed to the

Internet, as evidenced by Wu's next announcement. "We have twenty-two million officially registered Internet users in China," he says in late 2000. He also reports that the government-sponsored State Research Information Technology Company has hired a California company to help set up electronic commerce sites for thousands of government companies and a new wireless application protocol system that should greatly expand Internet access in a country where cellular telephones vastly outnumber PCs.

Not only are the number of users increasing, but in spite of Minister Wu, so is the content. The state media doesn't report that Li Jizhou, vice minister of public security, is currently in Beijing's Qincheng Prison awaiting trial for taking bribes worth millions of dollars, but the story is all over the Net. I once again try to reach foreign news sources as a test and discover that many sites in Chinese, English, and other languages are accessible from office buildings, homes, and an Internet café. Huang Qi is in jail, which cannot be forgotten, but www.6-4tianwang.com, his website, is up and running, carrying his message. In fact, there's an update on Huang's case. It says that he had been held in solitary confinement for two months, after which time he was seen being moved from one to another prison. A passerby reported that Huang shouted out a message: "Change the password on my computer! Don't worry about me." Someone apparently did change the password and continues to post new information about Huang's case and other human rights violations. There are many other examples. The government has recently ramped up its attack on Falun Gong, but the sect continues to fight back, using the Internet as its primary means of communication.

The most recent spike on Sina and other websites is caused by a controversial report about the 1989 Tiananmen Square massacre, which was based on documents that were purportedly smuggled out of China. *The Tiananmen Papers* exposes the Chinese leaders' squabbles over the decision to use violence in order to crush the protests. The government denounces the report, claiming that the documents are fakes aimed at destabilizing the country. "Any attempt to play up the matter again and disrupt China by the despicable means of fabricating materials and distorting facts will be futile," says Foreign Ministry spokesman Zhu Bangzao. However, Western scholars stand by the papers' authenticity.

Not surprisingly, the news sparks a vigorous online debate. Whereas censors silence some of the conversation (a message that details CNN's coverage of *The Tiananmen Papers* is reportedly deleted within minutes of appearing on Sina), a wide range of opinion survives on many websites. In fact, excerpts of *The Tiananmen Papers* themselves are posted along with students' comments about them on a Beijing University website that is accessible to anyone with a Web connection in China. It's the latest vividly profound evidence of the changes brought by the Internet. Now that *The Tiananmen Papers* is on the Net, distributed by e-mail and printed out, there's no stopping its spread or the spread of the accompanying debate. Paper cannot wrap up a fire.

CHAPTER 18

THE LONG MARCH TO CYBERSPACE

Bo and Heidi have big news: She is pregnant. They discuss the implications of another child coming at a time when Bo's commitment to be in China is increasing, and they resolve to buy an apartment in Shanghai and put Tiger in preschool there. Heidi seems as excited as Bo about the decision. As her Mandarin skills improve, she has become comfortable living in Shanghai and has begun to explore career options there. She plans to use her background in social welfare and project management to help organizations that are addressing such issues as rural education for young children.

When they go apartment hunting, Heidi and Bo are shown a brand-new building called Shanghai Famous People Mansion. In spite of the irresistible name, they choose a centrally located apartment in the Baroque Palace. Discreet from the outside, the entryway and lobby resemble a Roman brothel. It has white columns, gilded mirrors, and the cloying smell of perfume. When Heidi meets with a series of interior designers, she is presented with variations on what is apparently the hot Shanghai trend: American suburban chic. The decorators all propose a hot tub, his-and-her sinks in the bathroom, shag carpeting, and gaudy arrangements of fake flowers.

In December, Len Baker is back in Shanghai after a few days in Singapore, where he and Eric met potential investors. The Chengwei partners plan to take advantage of Len's stopover; they have asked him to reappraise a few of their portfolio companies. They arrange for new presentations by Oval, HDT, and a few others so that Len can evaluate their progress in the three months since the Chengwei conference.

This Oval presentation by John Sunn is a sharp improvement. He is polished and confident. In addition, he has impressive news to report. The Oval founders originally committed to forgo salaries until they signed their first customer. They can finally get paid after winning their first contract, an impressive deal with Goodbaby, the $1.2 billion company that produces 30 percent of the world's baby strollers and a large share of other baby products. To close the deal they were up for seventy-two straight hours, negotiating a $400,000 contract with the potential to grow to more than $4 million.

Next Len hears a presentation from the newly formed marketing company HDT, the result of the merger of T2 and iTom, before he, Bo, and Eric go to dinner. Len praises the progress at both companies. However, when he asks about IEI, Bo and Eric admit that they are discouraged by a stalemate over the hiring of a new CEO. Bo says that he doesn't know how hard to push. "I keep hammering them," he says, "but it isn't working."

"They are adopting a passive-aggressive stance," adds Eric. "They're rope-a-doping the process."

Len advises a different tack. "Put it on *them*," he says. "Say, 'It's true that we will own forty percent of the company, but you will own sixty percent. It's your company. If you don't want to bring in a CEO, let's not. At the same time, let's not pretend we're going to do it. If we pretend we're going to do it we're going to spend a lot of time and energy and it's going to fail. Let's move on. It's your call.' "

They drop him at his hotel, and Len assumes that Bo and Eric are heading home, but they instead call up the IEI founder, rousing him and asking for a meeting. When they face off at a bar, the man is wary of them after weeks of their badgering. Bo and Eric take turns trying Len's approach, but the IEI chief remains distrustful. However, by the time they finish at two, he looks at them differently. "We know we need a CEO," he says. "But we've been getting the message from you that we're useless. It sounds like you want a CEO to take our company away. We always wanted management help, but we have worked too hard to lose what we have created."

"You have been a significant business from scratch. That's why we're interested in going into business with you," says Eric. "However, we see you not as a ten million dollar business, but as a hundred million dollar business. It's why we have been pushing."

"We're learning how to communicate, and I admit we could do better," Bo says in an uncharacteristic conciliatory and quiet manner. He sits with his palms upward on the table. "Eric's right," he continues. "We wouldn't be here if we didn't believe in you. The question in our minds always has been a CEO and management team to supplement the company's strengths."

When the meeting ends, the three men shake on their partnership and agree to hire an executive search firm the next day.

Five hours later, Eric meets Len for breakfast. "By the way," he tells Baker. "We went out last night and thrashed things out about the IEI CEO. We're going to call in a search firm today."

Len shakes his head in disbelief. "This would have taken three weeks of meetings in the Silicon Valley," he says later. "In China, it's done by morning." It's an example of the ultra-speeded-up time frame of the IT movement in China, where years become months and months become days or hours.

Over the course of the thirty-six hours he spends in Shanghai, Len once again takes notes and fills in another chart in his small notebook. Afterward, he tears out the new chart and tapes it onto the other one. Comparing them, he notes the movement—"the type of movement you want to see." There are more companies in the plus column. Movement itself is expected, Baker says, but so much movement in a few months is more evidence of the compression of time in China.

Moore's law said that that speed of microprocessors would double every eighteen to twenty-four months; Netscape time described the six months it took for the premier Internet company to develop and release new versions of its software. Now there is Chinese Net time—faster and ever more intense. "It's dizzying," Len says.

The difference in his latest chart and the one he made in September is an indication of the strength of the Chengwei team and its entrepreneurs. Bo and Eric took two companies that had been in his negative column and, by combining them and recruiting a new CEO, created a strong company in the positive column. "It's an incredibly vibrant company with a roster of customers and a depth of products and services and an enormous market," Baker says. Oval had a good product and a potentially enormous market, but now it has a paying customer. More important, John Sunn and his management team have gained focus and confidence. "You see it when they walk in the door," Len continues. There's more: SeeWAP, with its impressive management and technical team but no customers, has metamorphosed into OneWave, moving its existing technologies and adapting them for one of China's most promising markets. In doing so, it has signed two of the most important regional telecommunications companies as clients. IEI, with great momentum, will now begin its search for a CEO. It, too, has the potential to become a powerhouse in an astoundingly huge market. It looks as if Chengwei will go ahead with the investment. There are other companies in the positive column, a few question marks, and three in the negative—in serious trouble. However, Baker sums up, "The chart was strong in January, but it's stronger now. I keep coming back to the most striking aspect: that the positive changes, in spite of the worsening economic climate, have occurred in three months." He continues, "Yes, I'm very happy with the portfolio, though it's ugly in a lot of dimensions. That's expected, however. It's expected because it's early. I've seen ugly and ambiguous and messy before; it's exactly where they should be."

When I sit down with Len, I ask him how much time Eric and Bo have to make successes of their companies and exit their investments. "The investors in this fund are patient," he says. "They don't expect overnight returns. Last year, there was a sense that it might be easy, particularly after the AsiaInfo IPO. It had a big run-up and everyone was seduced, but the fund was never set up with the idea of quick exits. The investors in Chengwei have the ability to take whatever time horizon is reasonable for the situation. If that is a year or two, great. If it's five or six or seven years, that's all right, too. Bo and Eric will get tired before the investors will."

Yes, it's a different climate, Len says, but the downturn in the market is "a positive" for Chengwei. He's not trying to spin things, either. "The bubble that we had last year caused a massive misallocation of capital and a massive misallocation of effort," he says. "Enormous amounts of emotional energy were going toward things that weren't fundamental. The economic climate has forced Eric and Bo to do exactly what great VCs must do. Their efforts are already showing up in the strength of their companies."

Eric, who joins the conversation, is vigorously nodding. "I admit that I'm feeling safer since the market tanked," he says. "I could never get my arms around what was going on the last couple years. I never understood it. I played a role in it and it was sort of exciting, but it was impossible to analyze. The tools I learned in school or in my business experience didn't apply. Now I'm comfortable again. My tool kit works. I'm able to sleep at night. At least a little."

When he departs China, Len is enthused. "Watching Eric and Bo learn is as exciting as anything," he says. "It's like Silicon Valley in the early days. We were reading the chapter a day ahead of the students. Nobody knew the formula because there was no formula. Same with Chengwei. More so, in fact. The conversations are much more direct and immediate in China. There's a stiffer emotional tone in the interchange. The difference is the sense of urgency. There's a sense that the fate of more than a billion people is on the line."

The three friends scatter. Len and Eric fly to San Francisco. In Shanghai, Bo and Heidi begin the celebration of the Year of the Snake. Lihui cooks a feast of fried duck, black-bean eggplant, wheat noodles with pork and mushrooms, and red sun dumplings. Heidi is busy editing for a local foundation supporting Chinese avant-garde art. She is also excited about meetings she has had with the Nature Conservancy, which is working with the Yunnan provincial government to develop a series of wilderness preserves and national parks across 16.5 million acres in the northwest section of the province. The Nature Conservancy is trying to get as much local support as possible, and Heidi and Bo have agreed to act as unofficial advisers for the project, helping to strategize and bring other business and government leaders onboard. They have already enlisted Edward, who has agreed to become a member of their Asia-Pacific Council.

Back in San Francisco, Eric and I meet at the Mandarin Oriental. Even before he takes his seat, he is ranting about the new schedule of the United nonstop from Shanghai, a disaster for road warriors. It leaves at noon, when it's impossible to sleep, he says, and arrives a dozen hours later at seven in the morning West Coast time, the start of a new day. This morning Eric touched down and raced to meetings in the Valley. It's now four in the morning in China but one in the afternoon in California, and he is operating in the surreal state of jet lag and no sleep, exacerbated by the barrage of calls from the press after Chengwei wound up in today's *New York Times*. Of all things, the company unwittingly found itself in the center of the confirmation hearings for Donald Rumsfeld, nominated for U.S. secretary of defense by the newly elected president, George W. Bush, who took the oath of office last week. Rumsfeld was grilled about his personal investments, including one in Chengwei. How does the investment fit with Rumsfeld's supposed hard-line position on China? Is it a conflict of interest?

Eric is annoyed that Chengwei is in the middle of the story, since it sounds as if something shady is going on. Plus the *Times* got it all wrong, naming Bo and his brother Tao as Chengwei's founders. Eric and Bo are concerned by the erroneous impression that their independent company is, like Tao's, tied to the Chinese government. Rumsfeld meanwhile promises to sell his Chengwei stake. Apparently to prove that he's no friend of China, in the hearings Rumsfeld attacks what he describes as the Clinton administration's "self-delusion" in thinking of the Chinese as strategic partners. He notes that the Bush administration will redefine China as a "competitive partner."

Hurst Lin must get a kick out of a new Accenture advertisement. (Accenture is the new name for Andersen Consulting.) A few years ago, Lin told me that one reason for founding Sinanet, now Sina, was his worry about the survival of the Chinese language. "Why should the computer world be dominated by English?" he asked. The Accenture ad, a full page in the *New York Times* and the *Wall Street Journal*, shows a newspaper headline that announces, "CHINESE TO BECOME NUMBER 1 WEB LANGUAGE BY 2007." In smaller type, it reads, "Now it gets interesting." Not only will Chinese be the number one language on the Web by 2007, but new official predictions, released by the government, say that China will be have more Internet users than anywhere, bar none, by 2003 or 2004. Early in 2001, in a particularly optimistic mood, Edward predicted there would be 300 million Internet users in five years. Now an analyst says that the number is realistic, though not until 2007.

Meanwhile there's a new, sober tone in Zhongguancun. Zhidong has what he thought he wanted. Sina is once again his company—on his shoulders. The company has continued to grow in terms of the number of devoted users and the depth and breadth of its website. There are 16 million registered users and some new and highly successful offerings on the portal, including SinaMail, which is similar to Yahoo Mail and Hotmail. There are already more than 6 million accounts—thirty thousand new ones a day. There's a new wireless version of Sina for Palm Pilots and other handheld devices. Revenue is down, however. At the first sign of an economic downturn, mainstream advertisers abandoned the Web. Even Yahoo, the world's online powerhouse, is struggling. The assumption that content would attract eyeballs and that eyeballs would bring dollars has been discredited, particularly now that there's increased talk of a recession in America that could spread around the globe. Sina is in dire straights. Its revenue sources are drying up, and its stock has sunk to the dangerously low price of $1 or $2. The original SRS RichWin software program isn't much help, either. Sina has released RichWin for the Internet and Sina 2000, which includes a Chinese browser, instant-messaging software, and a Chinese url system. However, many of the features are mirrored in Microsoft's Chinese products, which are greatly improved. The Redmond giant eventually gets at least some things right, and Chinese Windows is an example. It now incorporates what Zhidong once called "Chinese Mind." That is, the Microsoft program, created by Microsoft's Chinese team for Chinese users, incorporates the "feel" as well as the features of RichWin, including a sophisticated pinyin system, Chinese dictionary, and many other Chinese-language offerings. As a result, RichWin is becoming redundant. This turn of events has proved at least one thing:

Zhidong is not to become the Bill Gates of China. The Bill Gates of China will be Bill Gates.

In spite of the stress, Zhidong looks ever younger sitting back in his office chair with his arms stretched high, his elbows akimbo, his head resting on his small hands. He is preparing to head out to the Dragon Club. Zhidong is forever gentle eyed and sincere. However, his amiability does have a weight and sadness that comes from a decision he has had to make. Sina has to make deep cuts in its workforce. There soon will be an announcement of a 20 percent reduction.

There's an added surrealistic quality to his life these days because of a television show called *Qian Shou,* or *Hold Your Hand,* which has become a sensation and as such is the bane of Zhidong's existence. As if things weren't strange enough. Approached by a Beijing television producer, Zhidong naively agreed to allow his office to be used as a location for the *Twin Peaks*–like series. It never dawned on him that the show's viewers would confuse the real Wang Zhidong with the fictional Zhong Rui, a high-tech software engineer who graduated from Beijing University and went on to develop a famous software program before becoming one of the leaders of the Chinese IT revolution. In a recent episode, Zhong Rui dumped his wife for a girl in his office, an event that has divided much of China. One faction is furious at Zhong Rui for leaving his wife. The other cheers him for following his heart. Zhidong's real wife, Liu Bing, newborn double double dragon twins in arm (they were born in January), is teased wherever she goes.

The Dragon Club at the Longyuan Hotel isn't selected because of its charm. It's out of town and there are few distractions. A place to contemplate and plan. Zhidong is once again sequestering his generals, including Daniel Mao, Wang Yan, Yan Yanchou, and Hurst Lin. The last time they were here was before they worked with Hurst—that is, when Zhidong decided on the SRS-Sinanet merger.

The assembled group ponders its options. Sina could acquire Netease or Sohu at fire-sale prices, but such a move, though it would consolidate China's Internet portals, wouldn't solve the basic problems: revenue and the company's stock price. Zhidong can't help but feel discouraged, even though he is no less committed to Sina than at any time in the past. He states that he will do whatever it takes so that his website will survive, but the meeting ends without a resolution.

In early June 2001, Bo and Heidi, who are in California (their baby is due in July), get a startling call from Beijing. Liu Bing is calling to inform them of an alarming turn of events at Sina. Daniel Mao has orchestrated a coup—a successful one. When I spoke to Daniel a week earlier, he said that he was 100 percent supportive of Zhidong. Now Bing reports that Mao has maneuvered to sell the company in spite of Zhidong's opposition. The most likely suitor is China.com. Such a sale would be a crushing defeat to Zhidong, but the deciding factor may turn out to be China.com's well-timed IPO. The company has nearly $500 million in cash. If Sina had gone public on the original timetable, it would probably be the one buying the other players.

Liu cannot believe what has happened. Zhidong is fired. Daniel Mao is CEO. Liu, who Bo calls Zhidong's "human interface," says that her husband's love for computers is based on their dependability. "If you press A, an A will appear on the screen," she says. Humans don't work that way. Zhidong cannot conceive that Daniel Mao, his "friend and confidant," worked behind his back to take over his company. What's next for Zhidong? It's unknown. So is Sina's future. If the deal goes through, China.com's cash may help Sina survive in the current climate.

There are no guarantees, however. The ousting of Zhidong is immediately followed by the news that AOL Time Warner, the world's largest ISP, has entered into a $200 million joint venture with Liu Chuanzhi's Legend Computer. The two companies are each investing $100 million and will be equally represented on the board of the new company, but Legend will own 51 percent of the venture in order to satisfy the regulations that limit foreign ownership of Chinese Internet companies. Sina, whether it remains independent or it is owned by China.com, has a head start, but can it compete with the huge bankroll and expertise of AOL? Zhidong once said that the Chinese Internet must be dominated by a homegrown company. Legend qualifies, but AOL? After Zhidong's struggle, could the nation's most popular portal become one with initials that include an "A" for America?

The agreement initially is limited to AOL's consultation with Legend on its portal, FM365.com. In addition, AOL executives have promised to keep their hands off the content on a joint site if it is launched (perhaps in a year). However, the long-term view looks different. At some point the rest of AOL's users in America and across the globe—plus the company's other offerings, including the news—may well begin to leak into its Chinese sister company's website. For that reason, the implications of a mighty AOL-owned portal in China go far beyond the image of millions of Chinese turning on their PCs in the morning in order to hear the greeting *ni you xing jian* ("You've got mail").

Zhidong's shock over the firing wears off, and at a conference in Kunming on the weekend of June 16 he tells Bo that he is not giving up. In order for Sina to go public, the board approved the unprecedented arrangement that split the Chinese company off into a separate entity. That is, the company's most valuable asset was put into a domestic company and it is owned by two people. Sina China is owned by Wang Yan, who controls 30 percent, and Wang Zhidong, with 70 percent.

"Sina's listing prospectus doesn't raise the prospect of any transfer of shares upon Mr. Wang's departure," writes Leslie Chang in the *Wall Street Journal*. "It only notes that the company's structure would be viewed as entrenching his management position or transferring certain value to him, especially if any conflict arose with him." Daniel Mao tells Chang, "We are definitely doing the transferring at the moment and we don't foresee any legal problems." It's wishful thinking. Zhidong informs Bo that he has hired a team of lawyers.

When Edward arrives in San Francisco in January 2001, he picks up Stephanie from school and heads to the Mill Valley, California, house to play with her and Frances, fighting the urge to fall asleep. It seems as if it has been five months since he has slept. A year. Years. When the children are in bed, he and Jean talk. It's a breakthrough conversation. Jean agrees that the long separations aren't working for their family. She understands and respects Edward's commitment to China, so she agrees to try living there. It's an enormous sacrifice for her. Stephanie loves school and Jean loves America. But she says she will try.

Edward leaves for Phoenix, Arizona, for a Salomon Smith Barney telecommunications industry conference. At four in the morning on the ninth, the telephone rings in his hotel room. One of his managers in Beijing is calling. Edward has to return right away. Zhu Rongji's office has called. The premier wants to visit CNC on the day after tomorrow.

It would be a major event if the president of the United States were coming to an American company's office, but a visit by the premier may be bigger in China, where the nation's top leaders are far more imperious and inaccessible. Premier Zhu was instrumental in the government's decision to found CNC, but Edward never expected to meet him.

Edward calls Chareleson Zheng, who is with him in Phoenix, and asks for help getting back to China on time. Zheng comes to Edward's room, and the two use the hotel room's two lines to attempt to call travel agents and airlines. Almost every flight out of Arizona is booked for the next forty-eight hours, but they come up with an airline that offers him a route that could get him to Beijing just in time for Zhu's visit. Edward will fly at dawn from Phoenix to New York City, change planes for London, and then fly to Hong Kong before catching a flight to Beijing. He could easily miss the state visit, however, so Zheng calls the CNC office in China, which patches him through to a Beijing travel agent. The agent comes up with a China Eastern flight out of Los Angeles. "However," she says, "the airline doesn't take credit cards."

"How can an airline not take credit cards?"

By now it is six in the morning, and the flight leaves at eight.

Edward drains the ATM in the lobby, but it's not enough. Zheng and other friends who are staying in the hotel visit nearby ATMs and loan him the rest. With a fat stack of twenties in his breast pocket, Edward gets into Zheng's rental car. Inevitably, Zheng gets lost en route to the airport. They at last pull up to the terminal minutes before his flight will leave. Just then Edward's cell phone rings. It's Jean. She tells him that he left his passport in Mill Valley!

In a state of mild panic, Edward runs over to the ticket counter and pleads to get on a flight to San Francisco. Miraculously, he gets on the first flight out. Meanwhile, Jean books him on two flights out of San Francisco for Beijing, one on United and the other on Air China, which depart at noon and two o'clock, respectively. His San Francisco arrival is late—you can count on a late flight to or from SFO—and so he misses the noon flight. He touches down in San Francisco and finds Jean, who hands him his passport and ticket for the two o'clock flight. Edward hurries to the international terminal. He makes it. By seconds.

It is Tuesday afternoon, which means he will arrive in Beijing on Wednesday evening. He has the eleven-hour flight and then all night in China to prepare for the premier's visit. He writes a speech on the flight and then tears it up. He writes it again. He tears it up again. He writes it again.

A dozen CNC managers meet Edward at the airport in Beijing and whisk him to the office. He barks orders to his team. Informed that Zhu will be at CNC for forty minutes, Edward choreographs the visit down to the second. He and his team work throughout the night. Finally, before dawn, Edward showers and changes and then waits in his office, rehearsing the presentation one more time. When he looks out onto street, he sees that the roads are closed as far as he can see up and down the Xicheng District. In the distance, he sees the beginning of the motorcade with blocks of motorcycled police. He walks through CNC halls to make certain that the company is ready.

Premier Zhu arrives with a large entourage that includes a couple of familiar faces: Minister Wu Jichuan and the CAS's Hou Ziqiang. After introductions, Edward, remembering the forty minutes, leads Zhu and the others to CNC's demonstration room, where he jumps into the prepared talk. He demonstrates a live videoconference between him and a CNC staffer in Hangzhou, shows a sample of real-time IP-based television piped over the CNC network, and embarks on an explanation of the limitless potential of bandwidth, his favorite subject. Comparing his "broadband dream" to the dream brought by the people who introduced electricity to China, Edward says, "Premier, the difference is that China was behind the rest of the world when electricity was brought into the country, and we were slow to adapt and accept the technology. In fact, many of our people never took full advantage of it and the entire country suffered. But sir, we now have a chance to advance the dream of your generation. This time we are as advanced as any nation in the world. China no longer needs to follow, but can lead. This technology can transform the lives of our people. It can help to educate our people. It can bring opportunities for their lives. Please help us. Together we can achieve the realization of our broadband dream."

Zhu is quiet, intently listening, signaling that Edward should continue, but a security guard approaches Edward and hands him a slip of paper with scribbled writing. It says that the meeting must end. His time is up. Edward ignores it. Everything he has said thus far was meant to prepare Zhu for his pitch.

"Premeir Zhu," he says. "You can help this dream become realized in China by clearing the way for three things to happen." First, he asks Zhu to reform the regulatory environment. Second, he asks the premier to approve a license for CNC to be able to offer local access for broadband so that neither provincial nor local governments nor China Telecom can block the progress. Finally, he requests permission to send multimedia information over the CNC network—that is, permission for CNC to enter the content business, which is currently restricted.

A guard hands Edward another note that says, "The meeting must end now," which Edward ignores, too. He keeps talking until one of Zhu's escorts stands and interrupts, thanking Edward for the presentation. Zhu stands, too. He seems moved. Approaching Edward, Zhu says, "I am very impressed by your work here. You have accomplished a lot and done so very quickly. You have very good technology and assembled an impressive management team. I know that your work is very important to China, and I will do what I can to help you. I understand that you need changes in the regulatory environment. We want to make these changes so that you can continue your work. We're getting rid of the government's strangling influence on commercial ventures, but we're not doing it fast enough." He looks toward Wu and Hou and back to Edward. "I think we can do what you need."

The premier extends his hand. "Thank you," he says. "I wish I could stay longer. I hope we have the opportunity to meet again soon." With that, he and his entourage leave. When they're gone, Edward breathes heavily as his staff cheers. Edward looks out the window and watches the motorcade speed down the street. It takes a half hour for traffic in the financial district to return to normal.

Nearly hallucinating from exhaustion and relief, Edward leaves CNC and jumps in a car that drives him to the Chuiwei District, where he gets out in front of his parents' apartment building. He takes the elevator upstairs and knocks on the door. Liu Shu and Tian Yuzhao rush to the door. Both parents have tears in their eyes. "What happened?" "How did it go?" "Did you meet the premier?" "What did he say?"

He answers them. They want every detail. "And then what did you say?" "Then what?"

Tian Yuzhao is emotional. "Your mother couldn't sleep last night knowing that our son would sit with the premier. You have given us a great gift."

When Edward leaves to return to his apartment for some sleep, Tian Yuzhao slips a letter into his briefcase.

Weeks later, Edward is in Beijing, packing again. This time he is bound for Davos, Switzerland, for the World Economic Forum. He has been asked to be on two panels, one of them with Senator Orrin Hatch, Sony chairman Nobuyuki Idei, and Sun chief technology officer Bill Joy.

On the flight, he finds the letter in his briefcase. It is written in Yuzhao's tiny script, meticulously printed characters on both sides of rice paper. "I am very proud of you, my son," writes Yuzhao. "Your childhood was very difficult and you saw China at its worst. Instead of hating China and leaving our country forever, you returned to try to make it a better place for all of our people. I respect this decision."

Edward realizes there are tears in his eyes.

IT IS BETTER TO LIGHT A CANDLE THAN
TO CURSE THE DARKNESS

A four-tiered cruise ship glides down the Yangtze River. Bo and Eric are on board along with Heidi, Tiger, Eric's girlfriend, Yijing Zhu, Yangdong Shao, and Len Baker. It has taken Len, the Californian, to get the others on their nation's famous river. He's a nature lover with a passion for water. His life's ambition is to visit every one of Earth's rivers. It's impossible not to catch his enthusiasm.

They are on a cruise of China's famous Three Gorges, which remind Len of the Green River in the United States—only there's a striking difference. Whereas the wild Green runs through a five-thousand-foot-deep Utah canyon that is cut out of the desolate Tavaputs Plateau, the Yangtze, at least in parts, is bustling. Commercial boats chug upstream. Some riverbanks are populated, noisy, and polluted. Strange and surreal, there are places where the river meanders past two levels of civilization—villages at the water's edge and, several hundred feet above them, concrete apartment blocks. The explanation: The developments at the higher level will become the homes for people when the river is dammed and their villages are drowned.

The Yangtze courses from its source on the plateau of Tibet 3,900 miles to the East China Sea. In Chinese literature, the river has been described as China's spine and its blood and its heart. It has long transported the nation's people, manufactured goods, and crops on its roiling back. Millions live in villages and cities along its fertile banks. Now all anyone talks about on the Chang Jiang, which is the river's common name, is the dam being built a thousand or so miles upstream from the mouth near Shanghai. The dam will create a 400-mile-long lake over much of the 125-mile-long Sanxia, or Three Gorges.

The gorges, at the foot of misty and treacherous mountains, are magnificent. The precipices of the first, Qutang, form a gateway over the surging river below. In the second gorge, Wu, it's said that a clever eye can spot the deity who lives at Goddess Peak. It's wider and calmer but equally spectacular. Finally, Xiling Gorge, below cascading cliffs, has rapids that rip and twist over submerged reefs. The three lead to smaller gorges with vivid names that conjure up old China: "Military Manual and Precious Sword"; "Ox's Liver and Horse's Lung."

The Three Gorges will be flooded in less than a decade, when what will be the largest dam in the world is completed in 2009. The estimates of the cost range wildly—from $15 to $100 billion dollars.

The party line—literally the ruling Chinese Communist Party line—is that the dam will cure numerous ills. It will save lives by controlling the unpredictable river, which regularly floods and kills many thousands of people along its banks. It will provide a safer and deeper channel for oceangoing ships. Most importantly, it will become the world's largest hydroelectric power generator for China's future. There are dueling opinions about the dam, however. Environmentalists are enraged about what they describe as the reckless destruction of one of the world's most remarkable free-flowing rivers for dubious gains. Some engineers maintain that the dam will do little to stop the human destruction by the flooding Yangtze, since the worst calamities have occurred on tributaries that will be unaffected by the dam. Critics also claim that a series of small dams would be cheaper, do far less environmental damage, provide more electricity, and be safer. (If the Three Gorges Dam were to burst, more than 100 million people would be in peril.) Finally, many people decry the uprooting of up to 2 million people, most of whom are in families that have lived on the banks of the Chang Jiang for a thousand or more years. The government plans to move them to new and, it promises, better housing, and to compensate them and provide them with arable land or new jobs, but some of the first relocations have been disastrous.

The cruise ship *Elaine* looks like a white wedding cake. The Chengwei group is on the third deck in small staterooms. On this rare if brief three-day vacation for the partners and their families, there's invariably some business talk, but the group is transfixed by the scenery and nearly overwhelmed by life along the river.

On a concrete-gray morning, the ship docks at a lazy port and the travelers board a flat wooden sampan that's piloted by three shirtless boatmen who stand and row up a meandering tributary. A series of sampans powered by large diesel engines zoom by, and the boatmen laugh at the irony. The speedboats are heading upstream to set up tables of souvenirs for the tourists who are coming up the slow way.

Bo and Eric strike up a conversation with the boatmen, though it's difficult to understand their dialect. These men are from a farming family. In fact, this is the first generation in nearly a thousand years to work in a different occupation. Five brothers acquired this boat and work sixty days a year piloting tourists, earning two and a half times more money than they made in a full year as farmers. The money means that two brothers will be moving to the city. "When the river is dammed, more tourists will come," one boatman says. It's a boon for their family that not only means new housing, but the opportunity to expand their business.

A boatman who joins the conversation says that he is getting his own home, the first of his life. Indeed, extended families are being broken up. Eric and Bo, who translate for Len, hear how grandparents, parents, and children who now live together are being given their own apartments. For some, particularly the elderly, it's a terrifying break with tradition. However, this strong-armed and clear-eyed man is eager to see what happens. *No in-laws!*

Another day they take a side trip up another Yangtze tributary. Boatmen use poles to push their sampan up some rocky and steep stretches, but old-style flatboats, large and weighed down by heavy cargo, are pulled upstream by "trackers." Fifteen or more muscular and weathered men literally drag the boats up by the thick ropes that are tied to the hull. The men are naked in the chest-deep water. They sing ancient songs—songs sung by their forebears, who were trackers, too.

Ashore, Bo and Eric talk to some of the trackers whose fathers, fathers' fathers, and fathers' fathers' fathers—ancestors back fifteen hundred years—have had the same job. "The dam!" says one of the men. "We are moving to the city. I have a job in a factory." He seems eager. He has been told that he will earn three times more a year—the equivalent of $1,200 instead of $400.

At another stop, Bo and Eric, with Len listening in, talk to men whose faces have been darkened by a lifetime in the sun. They describe the history of their long-suffering farming families. The worse times have been when the Yangtze floods, which occurs without warning every four or five to ten years. Some of the floods have wiped out their villages and washed away their crops. The worst floods have killed as much as half of their family. The survivors flee to higher ground, only to return to their land and begin anew when the waters recede. "Again and again," says a fire-eyed man who has only a few teeth and a thin strand of hair on his sunbaked head. "Our people are stuck to this land, but we have a new opportunity." He and other farmers anticipate moves to new and better housing and better farmland. Still others plan to get transitional subsidies, knowing that they will soon have to find a new way to make a living. They believe the government's promises even as they know this is not like the floods; this time they will leave forever. Many of the farmers are contemplating a move to cities. In many cases, they expect to be able to quadruple or quintuple their income, assuming there are enough jobs to go around. (So far, there aren't.)

Bo, Eric, and Len are surprised that the people seem sanguine. International organizations have condemned the resettlement plans. About half of the families being forced to move are urban and the other half rural. While all will be relocated, a group of farmers that have already been moved were settled in an area with steep and rocky land that is barely tillable. They protest, but as yet no one has responded to their complaints. Some workers who have been moved to new cities have indeed been placed in higher-paying jobs, but others are in limbo waiting.

The million or two people who are displaced by the dam represent a small number of Chinese who are moving to cities from the vast rural lands. Millions—in fact, tens of millions—are being drawn by the prospect of higher wages.

In the evening, when the group is gathered in a circle of plastic blue chairs on the deck, Eric says, "The lives of *millions* of people are radicalizing!" The river is rolling gently. "Like the back of a long dragon," says a crew member. They are attempting to digest all they've seen today. A seismic jolt. Eric continues, "All I've read is that the dam is about electricity, economics, and the environment. But no, it's about transforming the life of a civilization. *Overnight!* Whether or not you like the dam, you can't deny the magnitude of the change it brings."

They drink green tea as the sun disappears behind the red-rock canyon wall and the group is lost in thought. Bo, sitting forward on his chair and exhaling his cigarette, finally says, "We're talking about releasing an amazing amount of productive energy. Think about it! If only there are enough jobs . . ."

Eric asks, "Yes, and what happens with the release of all that productive energy?" Back and forth, he and Bo follow the thought. China is already the world's manufacturing center. In the West, one is hard pressed to find something *not* made in China. Look along the aisles of any American Wal-Mart. Soon the products will carry Chinese brand names. In fact, Konka TV is one of the first. China's competitive advantage as a manufacturer makes it a near certainty.

For awhile, every Western product seemed to be made in Japan, but that country had a tiny fraction of China's labor force. "China is the only scalable manufacturing center anywhere," Bo explains. Indeed, from the riverbanks and farms, an enormous wave of people will join the manufacturing economy, a virtually unlimited source of labor. Provided unemployment is managed, workers will earn more and spend more. It's part of a spiral of economic growth. Back when I had dinner at the China Club with Feng Zhijun, Bo's father said that China is on its way to establishing the stabilizing force of a middle class. I didn't get it, but now I do. The exodus from the impoverished villages and farms. The employment of the poor. Making stuff for their countrymen and then for the West. When the Chinese have greater spendable income, the huge numbers of consumers in China will create a nearly insatiable market. In a 2001 Morgan Stanley Dean Witter report, the company's chief economist, Stephen S. Roach, writes, "Yes, China is a huge export machine. But the scope of growth in its domestic markets dwarfs anything that lies beyond its borders." Huge domestic demand will drive down the cost of production. No country will be able to compete with China's exports.

"One question is whether Chinese manufacturers can scale," says Bo. "And that's where we come in—the information revolution meets the traditional world of manufacturing. It's where technology plays the transforming role."

Now Eric says, "We—Chengwei—are catalysts for that process to happen. We are at the core. The whole model presumes that Chinese companies will stay ahead post WTO. How? Cheap labor isn't enough. They will succeed globally only if they become as efficient as the world's best companies. They can't do that without technology."

Since Eric and Bo steered Chengwei's investments away from dotcoms, many of the companies in the portfolio are in the business of helping traditional Chinese companies become efficient global competitors. They see and I see how Bo and Eric have come full circle in a year. Their idealism isn't dampened, it's refined. A year ago they were building the Internet to bring a revolution of information and communication. It was completely *other*—representing a new era that trumped and made irrelevant everything that came before it. It had nothing in common with the smoky and sweaty world of manufacturing. Now they understand that information technology provides the evolutionary edge for the nation's economic transformation precisely because it will revolutionize the nation's industry.

One of the times I raised the issue of China's human rights record with Bo, he said, "It's important to remember that much of China is so poor that there is barely any food. In such a vast country with so much poverty, focusing on economics *is* human rights." He's right. It may be that a degree of prosperity is a precursor for Edward's, Bo's, and others' dreams of an enlightened and free people. The elite in China who will benefit in the shorter term will have an impact on the country's social and political life, to be sure, but the most far-reaching changes will come when China's billions have a higher standard of living: healthcare, education, nutrition, and housing. It's a redefined vision of the revolution. As Liu Chuanzhi, the chairman and CEO of Legend Computer, later sums, "No, democracy cannot be imposed, but China will evolve in that direction as education and culture modernize. The shift from the current centralized system will come logically but only when the country is ready."

The group quietly listens to the river sounds—a motorboat's hum, lapping water, and the cries of vendors on the river's edge. A golden eagle flies overhead.

Their cruise is nearly over. They know that it's unlikely that they will ever again see the Three Gorges in their wild state. The reason stands in front of them: They are standing before an almost unfathomable edifice. It is the new great wall of China. Six hundred feet high. A mile and a half long. The dam.

CHAPTER 20

PAPER CANNOT WRAP UP A FIRE

*This is, after all, a new society, so sooner or later reason
will prevail.*
—Mo Yan, *The Garlic Ballads*

On March 6, 2001, in the impoverished province of Jiangxi, an explosion at an elementary school claims the lives of forty-two children and teachers. China's newspapers and Internet sites report that the explosion is a result of an appalling child-labor scheme. For two years, third- and fourth-graders have been forced to install detonators in firecrackers that are sold by a group of teachers to supplement their salaries.

In two days, Premier Zhu Rongji denies the reports, claiming that no explosives were kept at the school. He says that the explosion is "the result of a deranged man." Typical in China, the same media that carried the original story instantly retracts it and parrots the party line. Atypical in China, however, the true story refuses to die.

In Internet forums and chat rooms, Chinese citizens express their outrage about what they perceive to be a whitewash. Evidence that the government is lying, including interviews with the victims and their families, are posted on the Net and then printed out and circulated in coffee shops, on university campuses, in office buildings. Some sites, including the government's, delete many of the comments. However, even the website of the *People's Daily* carries a link to obscure websites that continue to discuss the catastrophe and the practice of child labor at the Jiangxi school.

The next headline that comes from Beijing is nothing short of astonishing: "Beijing Backs Down on School Explosion Story." The article reads, "Responding to reports on the Internet and elsewhere, Premier Zhu Rongji apologized Thursday for an explosion that gutted an elementary school in rural China." The turnaround is completely unprecedented. Premier Zhu has apologized. Largely because of the Internet.

"There is no tradition of public apology by Chinese leaders. Quite the contrary," writes Joe Klein in the *New Yorker*. "There has always been a vaguely Confucian assumption of patriarchal infallibility. Even after Mao Zedong's insane crusades caused the death of millions of people, Deng Xiaoping famously calculated that Mao's actions had been seventy percent correct. This was seen as a breakthrough in candor."

In his statement, Zhu says that the government bears "unshirkable responsibility" for the imbroglio and he orders an investigation. "History can never be covered up," he says. "The investigation will continue until we really get the full picture." Apparently the Chinese expression is accurate. Paper cannot wrap up a fire.

Eight months after his arrest, the trial of Huang Qi begins in Chengdu in February 2001. Two hours before he is led into the courtroom, he slips a letter to a friend, who posts it on Huang's website. In his letter, Huang hurriedly recounts his arrest and describes the condition of his incarceration. "The inmates assigned to monitor me ordered me to sleep under the toilet," he writes. "Every night I was handcuffed and shackled, slept on the wet ground, tasted the urine . . . [I could] hardly bear the torture. I tried suicide, but failed."

No supporters and only two lawyers are allowed to enter the court on the icy February morning. Before the hearing begins, Huang Qi faints and is taken not to the hospital but back to prison. His trial is indefinitely postponed.

Huang's case coincides with what appears to be a stepped-up campaign against human rights activists. There is a difference this time. The dissidents who are arrested are U.S. citizens or legal residents. One is Li Shaomin, a forty-four-year-old Princeton-educated sociologist. There is little outcry in the United States. "Perhaps it is hard for Americans to identify with Mr. Li or his family, because he has a Chinese face and name and is a naturalized rather than native-born U.S. citizen," writes Claudia Rosett in the *Wall Street Journal*. Li isn't formally charged with anything—he is "under surveillance," detained. According to Rosett, "Li not only traveled often to China, but lectured there on subjects such as how China might benefit from the Internet. As recently as last year he was receiving favorable coverage on Chinese state television. Then the political wind shifted in Beijing." Two other ethnic-Chinese scholars, both U.S. residents, are also detained. One, Gao Zhan, is formally charged with spying, an action that typically leads to a closed trial and a long prison term. (Li, Gao, and several other arrested scholars are tried, found guilty, and sentenced to prison terms, but they are subsequently expelled from China.)

At a train station in the northeastern city of Changchun, police arrest a Chinese activist named Chi Shouzhu who printed out pro-democracy material from a website. Chi, a factory worker, had been released in June after serving a ten-year prison sentence for taking part in the 1989 student protests. Another activist, Yang Zili, a thirty-year-old software engineer whose home page—Yang Zili's Garden of Ideas—carries a denunciation of "the soul of communism," is arrested in Beijing on his way back from his grandmother's funeral. His wife, Lu Kun, a kindergarten teacher, is arrested, too. "I thought I was kidnapped by bandits or rascals," she writes after being released. "At this moment, my every nerve was saying the same message to me: I [am] innocent and should fear nothing. In the past my husband Yang Zili had told me that in a country without the rule of law, a lot of innocent people would be harmed. I had not taken these words seriously. Not until I myself became a real victim could I deeply understand it." Her dispatch continues, "I was detained for three days without any sound reason. At 7:30 P.M. on the night of March 15, I went out of the gate of the Beijing State Security Bureau Detention Center. I walked up to a big tree away from the gate and started to cry there. I sorrowed for nothing but my fate of living in such a country where people have no security. I cried that I didn't know when I would be able to see my husband again."

On April 2, 2001, an American EP-3 spy plane on a routine surveillance mission near China collides with and breaks a Chinese F-8 fighter jet in two. The Chinese pilot is lost and presumed dead. Damaged by the collision, the U.S. plane makes an emergency landing on Hainan Island in the South China Sea. The twenty-four-person American crew is detained. The Chinese and U.S. governments blame and threaten each other, and the incident quickly escalates into the most contentious confrontation since the embassy bombing in Belgrade. Whose fault was the midair collision? The answer is at the center of the disagreement, since both sides insist that the other was at fault. China, maintaining that the spy plane bumped the fighter and killed its pilot, insists on an apology. The United States defiantly refuses to apologize, claiming that the Chinese pilot's recklessness caused the crash.

Tensions mount on both sides. President Bush, in the first international test of his presidency, demands the immediate and safe return of the American crew. President Jiang reiterates that China expects nothing short of an apology. Both sides say that future relations between the two countries are at stake.

There are no stone-throwing protests like the ones that followed the embassy bombing, but the Internet offers a window to the mood in China—mostly seething. More than 85 percent of the fifteen thousand people responding to an online poll say that they believe that the collision was a result of a "deliberate provocation" by the United States. Even many Chinese who acknowledge that the crash may have been the Chinese pilot's fault strike out in chat rooms and on message boards at what one describes as "American arrogance" and the "implicit hostility" of the spy missions: "Would America tolerate spy planes off California or Alaska?" The surge of traffic on websites is unparalleled.

The governments continue to posture while negotiators hammer out a deal. On the eleventh day of the crisis, China agrees to accept a letter from the United States that says that America is "very sorry" about the incident. The twenty-four crew members are sent home. They are greeted as heroes with ceremonies and seas of yellow ribbons.

In the days following the release of the crew, the Chinese government "[goes] all out" to convince its people that it has won a moral victory by releasing the crew, according to the *New York Times*. In the *People's Daily*, a commentary says, "The struggle by the Chinese government and people against American hegemony has forced the United States government to change from its initial rude and unreasonable attitude." However, the chat rooms tell a different story. Many Chinese citizens feel that their government caved in. "Our government is too weak. We have lost face," appears briefly on a Sina message board. The website's censors remove the missive and others. However, while the censors are dogged on the main websites, smaller sites are flooded with opinions contemptuous of the release of the crew. They publish the full letter of the American statement of regret so citizens can decide for themselves if the United States sincerely apologized or patronized the people of China. According to the *Times* article, such public opinion expressed on the Internet may have an ironic effect. The uncensored Internet in this case may strengthen the xenophobic factions of the Chinese government. It's a poignant reminder about the inherent nature of free speech. It is unpredictable. As are the results. It will be ironic if the Net causes a widening rather than a bridging of the gap between the United States and China.

When I first visited China a few years ago, most Chinese I met seemed enamored by America. There was a flood of American products into China, but the citizenry seemed far more excited about importing American ideals—ideals they learned about because of the opening economic and social ties between our two nations. Since then, the embassy bombing and then the refusal to apologize for the spy plane incident inflamed the Chinese people. So did President Bush's promise to do "whatever it takes" to defend Taiwan if China attacks and Secretary of Defense Donald Rumsfeld's rush to redefine China as a "strategic competitor" instead of the far more amiable "strategic partner," the description during the Clinton years. Now the Chinese seem far less interested in anything imported from America—particularly our values. The dangerous implications go beyond bad public relations if the Bush administration's aggressive stance has strengthened the position of the conservatives in the Chinese government. Beijing is preparing for a change of the old guard leadership when the current heads of state step down over the course of the next few years. Hard-liners who are the most threatened by the free flow of bits and bytes on the Internet—the ones who have attempted to squash the Net—are gaining power, in part because of the nationalistic wave of anti-American sentiment sweeping China. I still don't think that the Beijing government would ever pull the plug on the Internet, but the controls could worsen rather than ease if the conservative factions in the government who are the most threatened by the Net wind up consolidating their power.

Edward worries deeply about the fragility of the relationship between our countries. Anti-U.S. sentiment in China means that it's tougher for him to move forward with his network. For the first time ever, he has been asked, "How can you work with Americans?"

"For Chinese, these days America seems evil," he says. "For Americans, these days China seems evil. There has always been tension, but there are far more reasons to work together than not." He points out that many Chinese have an underlying respect and admiration for the United States. America has never colonized China. His father is part of a generation of Chinese who recall that Americans liberated them from the Japanese occupation in World War II. An American marine saved Tian Yuzhao when a Japanese soldier sicced a vicious wolfhound on him when he was nine years old. "The foundation is there for a respectful relationship," says Edward. "It's a love-hate relationship, and any successful relationship has to handle the love and the hate." It leads back to the theme of his life: "It's why the importance of connecting people cannot be minimized. Express your anger toward us. Let us express our anger toward you. Then let's talk." What's at stake? An open China. Far more influence about issues such as human rights and the environment. Free exchange with one of the most extraordinary cultures on earth. Perhaps most crucial, a strong partner and ally.

If the extremely high stakes aren't already obvious, they are made undeniably and tragically apparent on September 11, 2001, when, at 8:48 in the morning on the east coast, a jetliner smashes into one of the twin towers of the World Trade Center in New York City. Minutes later, a second jetliner hits the other World Trade Center tower. While there's little doubt that this is an unprecedented terrorist attack, the perpetrator and purpose are unknown. Soon another jet goes down outside of Pittsburgh, and then a fourth in Washington, D.C.; this one hits the Pentagon, which becomes a fiery inferno. By now, the world is watching live on the twenty-four-hour news channels, and the unthinkable happens: one and then the other World Trade Center edifice—110-story-high landmarks, symbols of Manhattan and the United States itself—collapses.

Some four thousand people are dead in the meticulously planned attacks. When it is revealed that the terrorists responsible were tied to Saudi-born Islamic extremist Osama bin Laden, it surprises no one; bin Laden is already on the FBI Most Wanted list after being connected to terrorist attacks on U.S. embassies in Kenya and Tanzania, the millennium bombing plot, and the 2000 attack on the USS *Cole* in Yemen. Three weeks ago, he told a journalist he would mount an unprecedented attack on the United States.

It's evening in China. Those with access to television or the Internet are riveted to the news. Frantic transpacific telephone calls and Internet messages fly both ways. It's nearly impossible to get through to New York City by telephone, but the Net continues to operate. Those with friends or relatives in New York or Washington anxiously wait for responses to their e-mailed messages.

The reaction in China is largely shock and deep dismay, but on the Net there are some troubling anti-American expressions. Before the government intercedes with instructions to Sina and other Internet companies to censor them, angry and vengeful messages appear in chat rooms. "I'm glad that the USA was attacked," reads one. Another: "Airplanes? Why not atomic bombs." Recalling last year's spy plane incident, some postings refer to "U.S. arrogance and totalitarianism." In spite of the ban, messages condemning America continue to post. One reads, "[The attack is] retribution for the U.S. bombing of the Chinese embassy in Belgrade and for crashing into our fighter jet over the South China Sea."

Are these the xenophobic rants of extremists or do they reflect the view of the majority of Chinese? In China, as in the United States, reactionary groups assemble online—with one important difference. Only the educated and relatively privileged have the ability to take part in the online discussion in China. They can't all be written off as a lunatic fringe.

There is also an enormous outpouring of sympathy and support in China. Many Internet messages blast those Chinese who are celebrating the attack. In Beijing, Chinese citizens gather in an impromptu vigil in Tiananmen Square, and throughout China people flock to the U.S. embassy and other American institutions, where they place flowers and letters of condolence on the sidewalk. An elderly woman approaches an American friend in Beijing. Grasping his hands, she speaks rapidly and emotionally, tears flowing. He doesn't understand everything, but can translate the words, "My heart weeps for your people—for all people." Late that first day, an e-mailed note arrives from Edward Tian. "Dear Friends," he writes. "It is with great sorrow and astonishment that I learned about the horrible attacks to New York City and D.C. I sincerely hope that all of your friends, relatives, and co-workers are fine despite the incidents. . . . Let's all hope that life goes on for the survivors, and the cities ruined recover to be stronger and better."

The polarized opinions of Chinese citizens are an accurate snapshot of a precarious balance that could tip either way. Once more the question is whether China will be America's ally or its foe, and not merely in the war against terrorism. The answer will have much to say about the world in this new century.

Both America's and China's governments' handling of relations is of paramount importance, of course. Hard lines and bullying by either side increase hostilities, whereas détente and olive branches—respect— build trust. We are witnessing a historic change in China. For the first time, the public's view of the United States has been seen to influence the government and even determine policy. It has the potential to help decide whether our two nations will face the future as partners or antagonists. Engagement—communication, two-way access to information, open exchange of culture and trade—will temper the anti-American sentiment in China and therefore embolden the progressive factions in the government. And it works both ways. The more contact Americans have with the Chinese, the fewer age-old stereotypes and fears fester.

In the aftermath of the attacks on America, the Chinese government's reaction is swift, but neither as swift nor as emphatic as most other nations'. President Jiang, in a call to President Bush, promises to work with the United States to fight terrorism, but China's support is less than enthusiastic. China is itself concerned about potential terrorism by fundamentalist Muslim Uighur separatists, who are being trained in Afghan camps associated with bin Laden's organization, but Beijing also fears U.S. retaliation because of the implications for its own security. There is already a significant U.S. military presence at China's front door in Japan and Korea. If America attacks Afghanistan, where the ruling Taliban government harbors bin Laden and his terrorist organization, there will be U.S. troops along China's western back door, too.

With the new openness and communication brought by technology, China's citizens know more than ever. Beijing has long bolstered itself by its campaign against "U.S. imperialism," but in the future the government will have fewer opportunities to filter opinion and censor the truth. As a result, if the United States winds up being viewed more favorably, the leaders who are pushing for closer ties will be strengthened, and Beijing will be more likely to work with America as an ally. A Sino-U.S. partnership can do much to fight the world's ills, from terrorism and regional conflicts to poverty and health crises. On the other hand, a tenser relationship between the two massive nations could hamstring or, worse, destabilize the planet.

The future is anything but certain, but discouraging setbacks are followed by jolts forward. The forces in China who push for engagement were greatly empowered when the nation was awarded the 2008 Olympics. Admittance into the World Trade Organization is imminent. Such historic events will further open the lines of communication and thwart the regressive factions in Beijing. The Chinese security forces ramp up their campaigns of filtering and surveillance, but local hackers and foreign companies such as Safeweb succeed in circumventing them. The flow of news and conversation is increasing, and now, working with Safeweb, the U.S. International Broadcasting Bureau, the agency behind the Voice of America, launches new technology that will aid Chinese users in getting around censors. "We want to force the Chinese government to accept the pro-democracy consequences of the Internet," Stephen Hsu, chief executive of Safeweb, tells the *New York Times*. Updating the type of radio broadcasts it blasted into the Soviet

Union throughout the cold war, the agency is also streaming live and archived audio and text news throughout China. In addition, it's flooding Chinese e-mail addresses with a newsletter similar to *Dacankao*. Some of the material is blocked, but much gets through. Information is flowing through the taps—if not quite like water.

After the trip down the Yangtze, Len Baker flies to San Francisco and Eric to Shanghai. In Beijing, Bo has a series of meetings, then dinner with Edward. A new workweek begins. The recent economic and political shocks—some of the biggest setbacks yet—are behind them, and yet they know that more are on the horizon. I watch with admiration and no small degree of trepidation as China's wired warriors meet each crisis with a redoubling of their efforts. It's difficult to imagine that they can work harder, but they must and they do. It's impossible to imagine that their mission can become more urgent, but it is.

Clearly China has a long way to go before the carp has completed its leap through the dragon's gate, but Bo and his friends are correct to point out how far the nation has come even since the spring of 1989, when millions of students marched on Tiananmen Square and throughout China. Ironically, the students may have succeeded in a way that no one could ever have predicted. Because of the idealistic movement of 1989 and the government's brutal response on June 4, people like Bo and Edward felt compelled to come home to China to make a new kind of revolution. Like Len Baker's description of Chengwei's portfolio, their progress is messy and sloppy. But it is also undeniable. As a supporter of Huang Qi wrote on the jailed dissident's website, which remains up as this book goes to press, "Fire can never completely burn out the grass, as it will grow up next spring."

> *But Captain Nemo paid little attention; his mind appeared to be absorbed by one idea, and without taking the proffered hand of the engineer,—*
> *"Now, sir," said he, "now that you know my history, your judgement?"*

ACKNOWLEDGMENTS

I am grateful to Bo Feng and Heidi Van Horn for their friendship. My times with Bo in the United States and China are some of the most memorable of my life. Heidi helped immeasurably with her wisdom and knowledge. Their children, Tiger and Xiaoyu, bring great joy into the lives of our family. Bo's loving parents, Feng Zhijun and Dong Lihui, and his brother, Tao Feng, were enormously gracious and helpful.

I have learned from and been enlightened by Edward Tian since our first meeting in Beijing. Since then, he and Jean Kong and their children have become close friends. Edward's commitment and intellect are boundless and inspiring. I also have great admiration for his parents, Liu Shu and Tian Yuzhao, who provided their moving recollections and observations.

Eric Li offered his wit, insights, and humor, and his and Bo's comrades at Chengwei Ventures, particularly Ma Ying and Jeffrey Jiang, helped me to navigate China. Bo and Eric are fortunate to have Sandy Robertson and Len Baker as mentors and advisers, as I was to have them as sources, generous with their time and bountiful knowledge. Their passion and devotion are infectious.

Wang Zhidong was warm, candid, and welcoming, opening the doors of his company. I am deeply indebted to many others in China and the United States who shared their wisdom and experience and helped in other ways. They include James Ding, Agnes Wong, Lui Bing, Yan Yixun, Liu Chuanzhi, Yan Yanchou, Wang Yan, Julie Hale, Chareleson Zheng, Gao Limin, Elaine Wu, Mark Fagan, Diane Chen, Jay Chang, Duncan Clark, Yangdong Shao, John Sunn and his team at Oval, Lin Hai, Shen Baojun, Richard Long, Hurst Lin, Jack Hong, Ben Tsiang, Jim Sha, Jon Karp, Ellen Anderson, and Adrian Zackheim. I am also in great debt to the people in China who chose to remain anonymous. I have long admired Orville Schell's work as a great China scholar, but I didn't realize until writing this book that he is also a great and instinctive teacher. He has no idea how much our conversation emboldened me to tell this story.

My friends Mike Moritz, Steve Randall, Adrian Lurssen, Owen Edwards, Brad Wieners, and Spencer Reese gave expert suggestions after reading my manuscript. (Adrian also suggested the book's title.) In addition to offering advice based on his immense knowledge of China and the Internet, James Mulvenon corrected my pinyin. The former editor in chief of *Wired,* Katrina Heron, understood the importance of the Chinese IT revolution before almost anyone and assigned a series of articles that became the building blocks of this book. I would also like to thank current *Wired* editor in chief Chris Anderson and senior editor Jeff O'Brien and my other magazine editors, including Rich Karlgaard, now the publisher of *Forbes* magazine, who assigned my first article about Bo; Barry Golson, who has been a friend and partner for two decades; along with Steve Randall, Arthur Kretchmer at *Playboy;* Alex Heard at *Outside;* Wayne Lawson at *Vanity Fair;* Robert Friedman at *Fortune;* and Jann Wenner and Bob Love at *Rolling Stone.*

I cannot thank Binky Urban enough for her wise counsel and heroic advocacy. At ICM, I am also thankful to Richard Abate and Holly Martin. The valiant editor of this book, Joe Veltre, worked tirelessly with me and provided an unwavering sense of this story's promise. At HarperCollins, I am immensely grateful to Associate Publisher Carie Freimuth for her commitment to this book; Sarah Beam's professionalism and kindness were indispensable; Lisa Berkowitz and Kate Kazeniac devotedly spread the word. Our talented copy editor, Jim Gullickson, left his mark (marks) on every page, and Bill Ruoto created the elegant design.

Don Barbour was the consummate researcher and reader (as he is the consummate father-in-law). He wields scissors and pencil with love and wisdom. My profound thanks go to him and the rest of my family: Joan, Sumner, Debra, Mark, Jenny, Becca Rose, Barrett, Nancy, Susan, Don, Lucy, Steve, and Mark. I get by with more than a little help from my family and friends, including Armistead Maupin, Susan Andrews, Buddy Rhodes, Terry Anderson, Peggy Knickerbocker, our MCDS friends, and the entire P.R.C. (Point Reyes Crew). Our dear friends Jennifer Van Horn and Jim Gislason contributed their stories, and Sarah Duncan, Michael Duncan, Merel Kennedy, and Nick Sheff consulted on everything from the title to the book's design.

There would be no book without Karen Barbour. She, Daisy, Jasper, and Nick are the brightest lights in my universe. I learn the most valuable lessons from them. With them, this is a wondrous journey.

Iacocca, Lee, 43
IBM, 46, 60–61, 105, 150, 157, 203,
    237–38
IEI Technology, 221, 250, 253, 258–59
Indonesia, anti-Chinese violence in, 7–8
Information Industries Ministry (MII),
    Chinese:
  CNC and, 146, 148–49, 225
  Internet regulation and, 101–3, 111,
      121, 140, 146, 198–203, 248, 253
  Sina and, 140, 198–203
InfoSec, 182–85, 220, 245
Intel, 63, 120, 156, 238
Interactive Week, 148–49
International Data Corporation, 106
International Herald Tribune, 106
Internet, xiv–xvi, 5–10, 15–16, 22, 41,
    55, 67, 98, 101–12, 119–29,
    144–61, 182–88, 192, 224–25,
    279–88
  on anti-Chinese violence in
      Indonesia, 7–8
  AsiaInfo and, 71–76, 78–81, 94–95,
      144, 151
  bandwidth costs and, 159
  BDI and, 70
  Chengwei and, 182–86, 194–95, 221,
      224, 244, 250–52, 276
  CNC and, 3, 145–61, 225, 229–32,
      248
  and collision of U.S. spy plane and
      Chinese fighter jet, 282–83, 285
  democratizing potential of, 103,
      106–10, 127, 129, 213, 283–84,
      287
  Ding and, 47
  on Falun Gong persecution, 8
  Feng and, xii, xiv–xv, 10, 19–20, 78,
      80–81, 86, 90–91, 103–4, 106–7,
      114, 132, 134, 139, 165, 194–95,
      209–11, 221, 276
  government initiative on, 111,
      145–61
  government regulation of, 8–9, 19–20,
      63–64, 87–88, 94, 101–5, 107–11,
      119, 121, 123, 128, 140, 146, 178,
      198–203, 207–8, 211–13, 248,
      253–56, 283–85, 287–88

growth of, 10, 104–6, 111, 120, 146,
    211, 255, 263
human rights and, 281–82
Legend Computer and, 238, 265
Li and, 167, 209–10, 276
Oval and, 241–42
RFTA and, 137
risks of investing in, 123, 178, 185,
    194–95, 199, 210, 241
September 11, 2001, terrorist attacks
    and, 285–86
Shanghai Online and, 72–75
shopping on, 105–7, 114, 137
SRS and, 86–92, 114–15
tanking of companies associated with,
    206–7, 209–11, 242, 261
Tian and, 3, 5–6, 31, 43–47, 71–76,
    78, 80–81, 103, 105, 248, 263
Tiananmen Square massacre and,
    255–56
and U.S. bombing of Chinese
    embassy in Belgrade, 123–29,
    285
Wang and, 63–64, 89–90, 92,
    114–15, 122, 126–28, 248, 265
*see also specific Internet sites and
    companies*
Intranet, 94–95
IP 800 project, 229–30
IT Daily, 139
iTom, 240, 245, 251–52, 258, 260
Ivanhoe Mines, 218–19

Janeway, Bill, 153
Jiang, Jeffrey (Jiang Shaoqing), 232
  Chengwei and, 175, 180, 185, 217,
      222
  RFTA and, 135–37, 175
Jiang Mianheng, 218–19
  CNC and, 145–46, 150, 218, 230
Jiang Qing, 27
Jiangxi, explosion at elementary school
    in, 279–80
Jiang Yaping, 129
Jiang Zemin, 103, 106, 109, 145–46,
    156, 282
  Clinton's televised conversation with,
      127–28

Montgomery Securities, 133
Moore's law, 63, 260
Morgan Stanley Dean Witter, 14, 205,
    276
    Sina and, 208–9
Mosaic, 45, 121
Mo Yan, 279
Mulvenon, James, 104, 107, 112
Mysterious Island, The (Verne), 23,
    26–28, 226

NASDAQ:
    downward spiral of, 207, 209
    and IPO of AsiaInfo, 204
    and IPO of China.com, 139–40
National Computer Bureau, 61
National Research Institute, 60–61
NationsBank, 133
Nature Conservancy, 261–62
Netease, Netease.com, 110, 140,
    198–99, 210, 223, 264
Netscape Communications, 45, 54,
    74–76, 114, 116, 120–21, 260
New Margin, 218–20
New Yorker, 209–10, 280
New York Stock Exchange, 70
New York Times, 36, 104, 108–9, 129,
    234, 248–49, 253, 262–63, 283,
    287
Ng, Thomas, 4, 247
Nietzsche, Friedrich, 173
Nippon Paint, 240, 251–52
Novell, 46
Nudist on the Late Shift, The (Bronson),
    155

Olympic Games, 9, 114, 208, 212, 231,
    287
OneWave, 250, 260
Opium War, 57
Oracle, 121, 179, 187, 232
Orchid Asia Holdings, 163–67, 171,
    175, 182
Ordermyfood.com, 187–88
Oval Technology:
    Chengwei and, 187–89, 216, 258
    CNC and, 232–33
    management of, 240–42, 260

Pacific Council on International Policy,
    87
Palm, Palm Pilots, 15, 104, 108, 186,
    245–46, 263
PanAsia Resources, 218
Peng Peng, 147, 157, 160
Pentagon, terrorist attack on, 285–86
People's Daily, 8, 38, 128–29, 211, 280,
    283
Perot, Ross, 169–70, 172
Perot Systems, 169–70, 175
Pioneering Portfolio Management
    (Swensen), 175
Power Corporation of Canada, 178–79,
    244
Pudong Airport, 18

Qianlong, Emperor of China, 235
Qian Shou (Hold Your Hand), 264

Rafferty, Carol, 77, 206
Railways Ministry, Chinese, 145–48,
    150, 157
Rand Corporation, 104, 107
Reagan, Ronald, 29, 179–80
Rebound, 185–86, 252–53
Red Azalea (Min), 27
Red Flag Linux, 182, 220, 245
Red Guard, 25, 28, 32, 51, 218
Red Herring, 247
Renren, 182, 185–86, 216, 222,
    252–53
Reuters News Service, 32, 204, 254
RichWin, 64–66, 86, 89, 91, 114, 263
Roach, Stephen S., 276
Robertson, Feng Technology Associates
    (RFTA), 133–37, 171–73, 175,
    180, 182, 185
Robertson, Sandy:
    AsiaInfo and, 79, 81–82, 206
    Chengwei and, 173, 178–79, 246
    Feng's relationship with, 14, 52–56,
        62–63, 82, 96, 99, 132–33,
        172–73
    Internet and, 63–64, 79
    Orchid Asia and, 163
    RFTA and, 133–34, 171–73
    SRS and, 63–64

PHOTOGRAPH CREDITS

Photographs reprinted by kind permission of the photographers, who retain the copyright to their works.

Preface: Bo Feng, teahouse, San Francisco. (Photo by Jennifer Leigh Sauer)

Introduction: Workers subcontracted by China Netcom to lay cables for fiber optics, Hebei Province. (Courtesy of CNC)

Chapter 1: The Bund, Shanghai. (Photo by Heidi Van Horn)

Chapter 2: Shanghai. (Photo by Justin Chan)

Chapter 3: Tiananmen Square, Beijing, 1989. (Photo by © Peter Turnley/CORBIS)

Chapter 4: Public phones, Beijing. (Photo by Mark Leong/Matrix)

Chapter 5: Edward Tian, Beijing. (Photo by Mark Leong/Matrix)

Chapter 6: Bo Feng, Ocean Beach, San Francisco. (Photo by Thomas Heisner)

Chapter 7: Zhongguancun, Beijing. (Photo by Mark Leong/Matrix)

Chapter 8: Internet Café, Beijing. (Photo by Mark Leong/Matrix)

Chapter 9: Wang Zhidong (seated) and Wang Yan, Beijing. (Photo by Mark Leong/Matrix)

Chapter 10: Computer billboards, Beijing. (Photo by Mark Leong/Matrix)

Chapter 11: Edward Tian at CNC, Beijing. (Photo by Peter Lau)

Chapter 12: Eric Li (left) and Bo Feng, Shanghai. (Photo by Justin Chan)

Chapter 13: Beijing Train Station. (Photo by Mark Leong/Matrix)

Chapter 14: James Ding, Beijing. (Photo by Peter Lau)

Chapter 15: (From left) Tao Feng, Bo Feng, and Eric Li, Shanghai. (Photo by Justin Chan)

Chapter 16: The Bund, Shanghai. (Photo by Justin Chan)

Chapter 17: Pudong, Shanghai. (Photo by Justin Chan)

Chapter 18: Shanghai. (Photo by Justin Chan)

Chapter 19: The Three Gorges, Yangtse River. (Photo by © Dean Conger/CORBIS)

Chapter 20: Goddess of Democracy, Tiananmen Square, Beijing, 1998. (Photo by © Peter Turnley/CORBIS)